Social Protection Programs for Africa's Drylands

A WORLD BANK STUDY

Social Protection Programs for Africa's Drylands

Carlo del Ninno, Sarah Coll-Black, and Pierre Fallavier

Contents

Boxes

Figures

Map

Tables

Foreword

Drylands—defined here to include arid, semi-arid, and dry sub-humid zones—are at the core of Africa's development challenge. Drylands make up 43 percent of the region's land surface, account for 75 percent of the area used for agriculture, and are home to 50 percent of the population, including a disproportionate share of the poor. Due to complex factors, the economic, social, political, and environmental vulnerability in Africa's drylands is high and rising, jeopardizing the long-term livelihood prospects for hundreds of millions of people. Climate change, which is expected to increase the frequency and severity of extreme weather events, will exacerbate this challenge.

Most of the people living in the drylands depend on natural resource-based livelihood activities, such as herding and farming. The ability of these activities to provide stable and adequate incomes, however, has been eroding. Rapid population growth has put pressure on a deteriorating resource base and created conditions under which extreme weather events, unexpected spikes in global food and fuel prices, or other exogenous shocks can easily precipitate full-blown humanitarian crises and fuel violent social conflicts. Forced to address urgent short-term needs, many households have resorted to an array of unsustainable natural resource management practices, resulting in severe land degradation, water scarcity, and biodiversity loss.

African governments and the larger development community stand ready to tackle the challenges confronting dryland regions. But while political will is not lacking, important questions remain unanswered about how the task should be addressed. Do dryland environments contain sufficient resources to generate the food, employment, and income needed to support sustainable livelihoods for a fast-growing population? If not, can injections of external resources make up the deficit? Or is the carrying capacity of dryland environments so limited that out-migration should be encouraged as part of a comprehensive strategy to enhance resilience? And given the range of policy options, where should investments be focused, considering that there are many competing priorities?

To answer these questions, the World Bank teamed up with a large coalition of partners to prepare a study designed to contribute to the ongoing dialogue about measures to reduce the vulnerability and enhance the resilience of populations living in the drylands. Based on analysis of current and projected future drivers of vulnerability and resilience, the study identifies promising interventions, quantifies their likely costs and benefits, and describes the policy trade-offs that will need to be addressed when drylands development strategies are devised.

Sustainably developing the drylands and conferring resilience to the people living on them will require addressing a complex web of economic, social, political, and environmental vulnerabilities in Africa's drylands. Good adaptive responses have the potential to generate new and better opportunities for many people, cushion the losses for others, and smooth the transition for all. Implementation of these responses will require effective and visionary leadership at all levels from households to local organizations, national governments, and a coalition of development partners. This book, one of a series of books prepared in support of the main report, is intended to contribute to that effort.

Magda Lovei

Manager, Environment & Natural Resources Global Practice

World Bank Group

Acknowledgments

This book is one of a series of thematic books prepared for the study, "Confronting Drought in Africa's Drylands: Opportunities for Enhancing Resilience." The study, referred to here as the "Africa Drylands Study," was part of the Regional Studies Program of the World Bank Group Africa Region Vice Presidency and was carried out as a collaborative effort involving contributors from many organizations, working under the guidance of a team made up of staff from the World Bank Group, the United Nations Food and Agriculture Organization (FAO), and the Consultative Group for International Agricultural Research Program on Policies, Institutions, and Markets (CGIAR-PIM). Raffaello Cervigni and Michael Morris (World Bank Group) coordinated the Africa Drylands Study, working under the direction of Magda Lovei (World Bank Group).

This book, entitled "Social Protection Programs for Africa's Drylands," was prepared by Carlo del Ninno (World Bank Group), Sarah Coll-Black (World Bank Group), and Pierre Fallavier (UNICEF). It draws heavily on the findings of two background papers: "Social Protection in Sahel's drylands: Strengthening resilience and promoting opportunities while contributing to equitable growth" by Pierre Fallavier (UNICEF) and "Review of social protection programs and projects in the IGAD region" by Jeremy Lind and Sarah Kohnstamm (Institute for Development Studies).

Early drafts of this book were reviewed by Michael Morris and Raffaello Cervigni (World Bank Group). The authors also want to thank Kalanidhi Subbarao (World Bank Group), who provided comments and suggestions on early drafts of the manuscript, and the participants in seminars held at the World Bank. Vanthana Jayaraj and Amy Gautam (World Bank Group) assisted with the publication process.

Funding for the report was provided by the Adaptive Social Protection Program multi-donor trust fund, currently supported by the Department for International Development (DFID), and the World Bank Group Africa Regional Studies Program.

About the Authors

Carlo del Ninno is a senior economist in the Africa Unit of Social Protection and Labor Global Practice at the World Bank, working on safety net policies and programs. He is the Manager of the Sahel Adaptive Social Protection Program. Over the past 13 years, he has worked on analytical and operational issues on safety net programs covering several countries in South Asia and Sub-Saharan Africa. Before joining the World Bank, he worked on food security policy for the International Food Policy Research Institute in Bangladesh, and on poverty analysis in several countries for the Policy Research Division of the World Bank and Cornell University. He received a PhD from the University of Minnesota and has published on safety nets, food policy, and food security.

Sarah Coll-Black is a senior social protection specialist in the Africa Region of the World Bank. For the past 10 years, she has been working on the design, delivery, and analysis of safety nets, including linkages with disaster-risk management and risk financing, and youth employment in East Africa. She is currently leading the World Bank support to the Productive Safety Net Program in Ethiopia. She co-authored the World Bank's Social Protection Strategy for Africa and articles on Ethiopia's Productive Safety Net Program. Prior to joining the World Bank, she worked in the Philippines on efforts to extend services to the poorest and most marginalized. She holds an MPhil from the Institute of Development Studies, University of Sussex, and an Economics Degree from Dalhousie University.

Pierre Fallavier is a social scientist and planner with 20 years of experience in social development and humanitarian programs and policies in Asia and Africa, working with aid agencies, local governments, civil society, and academia. He designed, led, and evaluated development projects and responses to crises in post-conflict and fragile areas, formulated and assessed social protection, poverty reduction, and disaster-risk management initiatives, and led policy research. He holds a Master in Community Planning from the University of British Columbia, and a PhD in Urban Studies from MIT. He currently works as Chief of Social Policy, Planning, Monitoring and Evaluation for UNICEF South Sudan.

Executive Summary

Introduction

This book, one of several prepared as background pieces to the larger Africa Drylands Study, explores the role of social protection in promoting the well-being and prosperity of the populations residing in the drylands of Africa, with a specific focus on the Sahel and the Horn of Africa. It argues that social protection policies and programs, which aim to directly reduce chronic poverty and vulnerability to shocks, have an important role in promoting the resilience of the people residing in these areas, and it shows how social protection programs, when well designed and carefully implemented at scale, have the potential to reduce sensitivity to the impacts of shocks and promote positive coping capacity. However, given limitations in scale and scope of existing social protection systems in the region, many countries continue to respond to chronic poverty and predictable needs through the humanitarian appeal system.

Poverty, Vulnerability, and Need for Social Protection in Africa's Drylands

Despite a decade of economic growth, many people throughout parts of Africa remain vulnerable to shocks and chronic poverty. This is particularly the case in the dryland areas of Africa, where recurring episodes of drought, disease, price spikes, and civil conflict continue to cause widespread food insecurity and hunger. The breadth and depth of this vulnerability is strikingly evident across the region spanning from the Horn of Africa to the Sahel.

Aligning with the focus of the main study *Confronting Drought in Africa's Drylands: Opportunities for Enhancing Resilience* (Cervigni and Morris 2016), this book concentrates on a subset of people vulnerable to poverty, shocks, and the negative impacts of climate change. Drawing on the definitions and methodology laid out in the Africa Drylands Study, families and communities that are most exposed to the impacts of climate change are defined as those living in dryland areas, that is, classified by the Food and Agriculture Organization (FAO) as hyper-arid, arid, semi-arid, and dry sub-humid. Among them, we consider as "sensitive to the impacts of climate change and of extreme climate events" households whose livelihoods directly depend on agriculture, that is, those who are crop farmers, pastoralists, or agro-pastoralists. Finally, we focus on a subset of the sensitive group, defining households vulnerable or unable to cope with shocks as those with an income below the poverty line of US$1.25 per person per day in an average drought year, below US$1.44 per day during a year of mild drought, US$1.63 during a year of medium drought, and US$1.81 during a year of severe drought.

We also limit the scope of our analysis to a subset of Sub-Saharan African countries, those that have the highest shares of populations living in drylands. They are all partly situated on the southern part of the Sahara Desert, from Mauritania on the western coast to Somalia on the eastern tip of the Horn of Africa. To match the classification used in the larger Africa Drylands Study, we classified them in two groups: West Africa and East Africa. To estimate the population exposed, sensitive, and unable to cope with droughts in the drylands over time and ensure consistency of analytical approaches across the sections of the report, an "umbrella model" was developed as part of the larger Africa Drylands Study.

The proportion of the population exposed to shocks—that is, residing in the drylands—varies between the Sahel and Horn of Africa. This book shows that the population in the drylands is expected to decrease in the Horn of Africa and remain the same in the Sahel. The distribution of those sensitive to climate shocks will depend on differences of livelihoods across the drylands and on the projected evolution of these livelihoods in the face of economic growth. The population sensitive to shocks in the Sahel and the Horn of Africa is expected to increase in 2030 under medium estimates of gross domestic product (GDP) and population growth. Within this context, estimates suggest that without appropriate interventions, a large number of agriculture-dependent households in the drylands will remain poor—and hence vulnerable to shocks—even in a context of high economic growth. While poverty rates vary from country to country, in the Horn of Africa they are markedly higher among rural populations residing in the drylands, as compared with the national population. In a similar fashion, evidence suggests that in many countries, the rates of poverty tend to increase with the Aridity Index.

Conceptual Framework—Social Protection Reduces Vulnerability and Improves Capacity to Cope

Social protection systems, programs, and policies help individuals and societies to build resilience to risks, achieve equity, and avail themselves of opportunities. They do this by providing basic income support to the poor and minimizing the risk of households suffering from adverse events while enhancing opportunities for a better future.

- Governments and the international community have increasingly drawn upon social protection instruments to respond to disasters, both slow onset events, such as drought, and sudden crises, such as earthquakes. The experience from these initiatives has demonstrated that social protection systems that are in place before shocks hit are able to respond in a more timely and effective manner than are initiatives that need to be launched in response to the shock.

• At the same time, social protection policies and programs help households and communities build their resilience before shocks hit. This is achieved at the household level for instance through the regular distribution of cash transfers accompanied by training activities to help diversify livelihoods away from climate-dependent activities, and at the community level through public works that invest in the natural environment—with re-greening initiatives, or in disaster-resilient productive infrastructure, such as dikes or improved irrigation systems.

• Over the last decade, there has been growing attention to the role that social protection can play in responding to climate change. Viewing social protection through the lens of climate change focuses attention towards how social protection can contribute to adaptation via livelihood promotion and diversification and promoting access to new opportunities.

• In the Sahel, for example, a World Bank-implemented program was launched, with the support of Department for International Development (DFID), to help develop adaptive social protection systems consisting of a combination of policies and programs to help poor and vulnerable households reduce the impact of climatic change and other shocks to build household and community resilience, and foster access to income-earning opportunities.

Social Protection Programming in the Drylands

Coverage of social protection programs across the drylands of Africa is limited, even though it is often higher in dryland areas than in other areas, like in Kenya. In the Horn of Africa and the Sahel, most social protection programs are small, fragmented, and largely donor-driven. Still, countries such as Ethiopia, Kenya, and more recently Uganda have scaled up their investments in social protection with encouraging results, providing a model for how other countries can progressively expand coverage to poor, vulnerable populations. At the same time, countries in West Africa are exploring how to establish social protection systems that are designed to adapt to climate change and to flexibly respond to the needs of populations vulnerable to shocks.

Coverage of Social Protection

Social security, most commonly pensions for civil servants and employees in the formal, private sector, covers only a fraction of the population and often tends to benefit better-off segments of the population. In the Sahel, the highest proportion of the population covered by formal social security is at best 13 percent.

Safety nets were introduced in several countries as an alternative to the annual emergency interventions of food aid distribution. There has been a proliferation of safety nets since then, with many continuing to be emergency, short-term responses to chronic humanitarian needs. The coverage of safety net programs in the Sahel is more limited than in the Horn of Africa. Across the Horn of Africa, there are numerous examples of efforts to consolidate and expand safety net coverage. In the Sahel, most safety net programs remain on a pilot basis or are essentially temporary post-crisis interventions focusing on specific geographical areas.

Insurance and labor programs include some experimental pilots of insurance products (particularly those that are index-based on weather events) and a very limited number of labor programs, particularly targeting young people.

Traditional safety nets have played an important role in mitigating the negative impacts of shocks. However, these informal coping mechanisms themselves have largely been stretched to their limits in communities hit by covariate shocks (droughts, price hikes, etc.) that weakened everyone's abilities to support themselves and one another.

Spending on Safety Nets

Spending on safety net programs in the Horn of Africa and the Sahel is generally low, even in comparison to other countries in Africa. Within this general trend, enormous variations exist across countries, reflecting the different scales of coverage, varying payment levels for beneficiaries, different payment modalities (for example, cash and food), and administrative costs. In all countries, a significant proportion of funding for safety net programs is from donor agencies. In Niger, donor funding accounts for 70 percent of all expenditure on safety nets.

Capacity of National Social Protection Programs to Respond to Dryland Vulnerabilities

Several countries in East Africa have made efforts to tailor safety net programs to meet local needs. In Kenya, for example, the Hunger Safety Net Program was designed specifically to respond to the vulnerabilities of people living in the arid and semi-arid areas of the northern part of the country. The Hunger Safety Net Program uses cell phone-based technology to support a mobile payment system that is adaptable to pastoral livelihoods. In Ethiopia, under the Productive Safety Net Program, efforts have been made to tailor the design and delivery of assistance to the pastoral regions of Afar and Somali. Beyond cash transfer programs, however, few social protection interventions have specifically targeted the drylands.

Among African countries, only Ethiopia has established the capacity to expand the coverage of its safety net program rapidly in response to shocks. The scalability feature of the Productive Safety Net Program was designed to provide a first line of response to drought, complementing the existing humanitarian appeal mechanism, which will continue to be used to respond to needs in areas outside Productive Safety Net Program districts or in cases where needs within Productive Safety Net Program districts exceed available resources. During the 2011 Horn of Africa crisis, the administrative and logistical infrastructure of the

Productive Safety Net Program proved capable of scaling up the coverage of the program very rapidly, thereby strengthening the capacity of hundreds of thousands of vulnerable households to withstand a series of unexpected shocks.

Humanitarian Response as a Social Protection Instrument

Because social protection programs are generally very small—and because few have the capacity to scale up rapidly in response to shocks—most governments rely on food aid and on humanitarian appeals as a means of responding to chronic poverty and predictable emergencies. Humanitarian assistance in the drylands typically involves the provision of food, cash, and other in-kind resources and services to help affected households cope with the immediate effects of drought. Delivery mechanisms for humanitarian aid often consist of food distribution, cash transfer programs, feeding programs, purchase of livestock, and provision of health services and water and sanitation services.

Humanitarian assistance is an appropriate short-term response to emergencies, but in many countries it is provided year after year in the same areas and to the same recipients, suggesting it is being used as a long-term instrument to address chronic poverty. This use of humanitarian aid is inappropriate, because the delivery costs tend to be extremely high. In addition to being expensive, protracted use of humanitarian assistance is often ineffective. While emergency distribution of food can save lives, it faces implementation challenges. The food often arrives late and the amount delivered is generally less than what is required. Additionally, given the emergency nature of the support, it is often difficult to target the poorest and most vulnerable households; the authorities tend to focus mainly on getting the resources to communities that have been especially hard hit, but the allocation of resources to households within these communities is often done in an *ad hoc* fashion, or the resources are made available to all households, regardless of need. Finally, because humanitarian aid resources become available only after a shock has occurred and donors have had time to respond to appeals, the timing and amount of transfers received by the affected households tend to be inadequate to meet all of their needs.

Opportunity for Reducing Sensitivity and Improving Coping Capacity

Social protection programs, when correctly designed and effectively implemented, can reduce vulnerability in the drylands by reducing the sensitivity to shocks of vulnerable households and by improving the capacity of these households to cope after a shock has occurred.

Reducing Sensitivity

Social protection programs can reduce sensitivity to shocks by enabling poor and vulnerable households to invest in human capital, build assets, and diversify their livelihood strategies. The social protection programs that perform this function are those that target the chronic poor and provide continuous assistance over a sustained period. While the assistance is provided over multiple years, the expectation is that for individual households it is finite, in the sense that it will be suspended once the household has built an asset base and diversified its livelihood strategy, because at that point the household will be resilient and will no longer require support. These objectives are more effectively achieved when social protection support is combined with investments in human capital and livelihoods, and when it is integrated with other development programs such as those being proposed for dryland areas.

Cash transfers allow households to invest in human capital, build assets, and diversify their livelihood strategy. Increasingly, the delivery of financial support is complemented with other services, such as those that promote nutrition or provide skills training. This approach is becoming particularly common in the Sahel. These programs, when properly designed, can support more productive and potentially more diversified livelihoods and help participants to participate in the growth process and take advantage of the investments made in agriculture and pastoralist and other activities proposed in this report. A large and growing body of evidence shows that cash transfer programs work, including in dryland areas. In arid and semi-arid zones of northern Kenya, households receiving regular cash transfer support from the Hunger Safety Net Program withstood a severe drought in 2011 without any increase in poverty levels, whereas among those not receiving cash transfer support, 5.3 percent of households had fallen into the bottom income decile following the drought. In Ethiopia, the average period during which households participating in the Productive Safety Net Program reported being food secure increased from 8.4 months in 2006 to 10.1 months in 2012.

Public works can help households and communities reduce their sensitivity to shocks. In addition to delivering immediate assistance to participating households by paying wages, public works can put in place productive infrastructure that can improve permanently the livelihood strategies of recipients and their communities. Through public works initiatives, the Productive Safety Net Program has for instance helped construct 600,000 kilometers of soil and stone bunds that enhance water retention and reduce soil erosion. Public works initiatives supported under the Productive Safety Net Program have also been used to protect 644,000 hectares of land with area enclosures, leading to improved soil fertility and increased carbon sequestration. Within these enclosures, groundwater levels are rising, springs last longer into the dry season, and woody and herbaceous vegetation have increased.

Insurance programs can reduce sensitivity to shocks with initiatives that facilitate access to insurance products that lower the risk associated with traditional livelihood strategies, such as farming and livestock keeping. Over time as they become confident that insurance products can provide effective protection against the negative effects of shocks, households will be encouraged to invest in more productive livelihood strategies that will reduce their chances of falling into poverty.

Improving Coping Capacity

In addition to reducing sensitivity to shocks, social protection programs can improve coping capacity and help households recover after a shock has hit by providing immediate assistance, usually in the form of food or money. This second type of program—often referred to as "temporary" safety nets—is designed to provide short-term assistance to help affected households cope with the effects of a specific shock. This type of program allows households to avoid the use of short-run negative coping strategies that will undermine their livelihoods over the longer term, such as selling livestock or pulling their children out of school. However, although households that receive benefits through this type of program may avoid falling deeper into poverty, they will not necessarily be resilient in subsequent years, after the flow of benefits has stopped. It is critically important that whatever instruments are used be part of the permanent system and that they be rapidly scalable and coordinated with humanitarian support, so that humanitarian support can be mobilized quickly when the capacities of scalable safety net programs are exceeded.

Building Adaptive Social Protection Systems That Respond to the Needs of People in Drylands

To be effective, a national safety net program must be capable of rapidly scaling up the provision of transfers to people who have been (or will be) negatively affected by a shock. The best scalable safety nets are able to quickly respond to an imminent or emerging crisis on the basis of information generated through early warning systems and seasonal assessments. To date, however, few safety net programs in Africa have the capability to respond to shocks. Ethiopia's Productive Safety Net Program is at the forefront in this regard, while Kenya is building rapid response capacity into its National Safety Net Program, particularly within the Hunger Safety Net Program.

The success of these models hinges on a well-functioning early warning system, which forecasts a crisis or provides detail on the areas and people most affected and clear, objective "triggers" that determine when resources should be deployed. This mechanism is complemented with the creation of a unified registry of safety net beneficiaries and the contingency planning capacity of the social safety net program. Hence, in Kenya, the creation of a unified registry—together with a payment mechanism that covered all households in the program area—enabled the Hunger Safety Net Program to scale up quickly in response to drought. Where such systems do not exist, safety net programs can use existing mechanisms to identify and reach those in need of support. In many cases, humanitarian response programs offer insights and options into how this can be achieved.

More broadly, to strengthen resilience, social protection should be integrated with disaster risk reduction initiatives and emergency relief operations through a common understanding of the different types of vulnerabilities to shocks, better early warning and planning of coordinated early action, and development of capacity to quickly target and reach people affected by crises. In this way, social protection becomes part of a larger integrated system of risk management that links disaster risk reduction with risk-informed development programming and links prevention and development with humanitarian responses.

At the same time, lack of government capacity is a real constraint to extending the coverage of existing social protection programs. Such capacity limitations are particularly acute in remote pastoral areas. Recent innovations in delivery mechanisms, particularly the use of information and communications technology (ICT), offer opportunities to reach remote populations, which is of particular interest to dryland regions. In northern Kenya for instance, investments in solar panels and smart card technology have enabled the Hunger Safety Net Program to create a payment system that is responsive to the mobile lifestyles of pastoral populations.

Financing a National Adaptive Social Protection System

Among African governments, limited financing has long been seen as an impediment to developing integrated systems of social protection, which are perceived as a luxury that poor countries cannot afford. However, international evidence suggests that it is possible to achieve national coverage for a target population with 1–2 percent of GDP. The annual cost of the Productive Safety Net Program in Ethiopia is equivalent to 1.2 percent of GDP, while safety net coverage in Kenya is equivalent to 0.80 percent of GDP. The efforts of these two leaders can provide examples to many of the Sahelian countries, as well as to Somalia, Sudan, and South Sudan, whose investments in safety net programs have been modest. With Senegal spending 0.27 percent of its GDP on safety nets, Mali 0.5 percent, and Burkina Faso 0.6 percent, only Mauritania with 1.3 percent is in the 1–2 percent range of the average expenses in poor nations globally.

While safety net programs are often thought to be expensive, studies and simulations have shown otherwise: in Niger, providing regular cash transfers to support 20 percent of the poor population would cost US$102 million per year, as compared with an annual average of US$218 million spent on humanitarian response, on average, between 2010 and 2013.

Costing of Safety Net Coverage in Africa's Drylands

National safety net programs may be cost-effective relative to humanitarian responses, but they can still require a significant commitment of resources—with the size of the commitment depending on the scope of coverage and the level of support provided. The umbrella model was used to estimate the number of vulnerable people in 2030 who could be made resilient in the Horn of Africa and the Sahel using some conservative cost estimates. The estimated cost in 2030 of providing safety net support to bring all drought-vulnerable households to the poverty line ranges from less than 0.5 percent of GDP in countries with relatively high GDP per capita (for example, Mauritania) to almost 5 percent of GDP in countries with relatively low GDP per capita and extensive dryland populations (for example, Niger).

The umbrella model was also used to estimate the cost of a basic scalable social protection program that can cover 25 percent of the poor and vulnerable population in a normal year, or 35 percent of the vulnerable in case of a mild drought, 50 percent for a moderate drought, and 65 percent for a severe drought. In West and East Africa, covering 25 percent of the estimated 101 million vulnerable in a regular year in 2030 would cost 0.3 percent of GDP. In case of mild, moderate, and severe drought the cost would go up to 0.5, 0.76, and 1 percent of GDP, respectively. It is clear that poorer and more vulnerable countries like Niger will need a proportionally larger amount of resources and therefore may be more in need of foreign aid, especially in a drought situation.

The results of the model show that a minimum level of social safety nets—which is essential to protect the poorest from chronic stress and from sudden climate-related shocks—is affordable. If well designed and efficiently administered, they can cover a reasonable percentage of the poor population and make them more resilient to droughts. Social safety nets are a low-cost approach to not only help bridge a revenue gap for the poorest, but also impart them with human and physical capital to strengthen their livelihoods against crises (through the training that comes along with the cash transfers, the climate-resilient productive infrastructure built through cash-for-work, and the incentive to diversify from the agriculture-insurance component).

The challenge for policy makers is to strike an appropriate balance between permanent and emergency programs. However, the emerging experience with scalable safety nets suggests that investments made in permanent systems reduce the costs associated with delivering support to households negatively affected by the impacts of drought.

Conclusion

Social protection programs will be a key component of strategies to increase resilience and reduce vulnerability in the drylands. If present trends continue, by 2030 the dryland regions of East and West Africa will be home to an estimated 429 million people, up to 24 percent of whom will be living in chronic poverty. Many others will depend on livelihood strategies that are sensitive to the shocks that will hit the region with increasing frequency and severity, making them vulnerable to falling into transient poverty. Social protection programs thus will be needed in the drylands to provide support to those unable to meet their basic needs.

Safety net programs can increase resilience in the short term by improving the coping capacity of vulnerable households. Rapidly scalable safety nets that provide cash, food, or other resources to shock-affected households can allow them to recover from unexpected shocks. Scaling up an existing safety net program can be far less expensive than relying on appeals for humanitarian assistance to meet urgent needs. Despite the fact that safety nets are a more effective response to poverty and vulnerability than emergency assistance, funding for safety nets is low and flows of humanitarian resources to countries in the Horn of Africa and the Sahel remain high.

Social protection programs can increase resilience over the longer term by reducing sensitivity to shocks of vulnerable households especially if combined with other development programs. Safety net programs must be complemented by other types of social protection programs that enable chronically poor households to build their productive assets and expand their income-earning opportunities. Providing predictable support to chronically poor households and enabling them to invest in productive assets, as well improved access to basic social services, can effectively reduce their sensitivity to future shocks and help them to participate in the growth process and take advantage of the investments made in agriculture and pastoralist activities proposed in the drylands.

The dynamic nature of vulnerability in dryland areas draws attention to the need for safety net programs to be able to scale up in the face of shocks and then to scale back down when these pass. In dryland areas, such instruments may be even more important than in non-dryland areas given the levels of vulnerability and exposure to shocks. Emergency support should be provided on an occasional basis whenever a set of pre-defined triggers are met and in a manner that complements, rather than replaces, the support extended through scalable safety nets. Effective early warning and monitoring systems are needed to alert policy makers and guide the response.

Social protection programs must be tailored to address the unique circumstances of dryland populations. Program delivery mechanisms similarly need to respond to the specific needs of dryland populations (for example, by accommodating the mobility of pastoral populations).

Capacity constraints will need to be overcome to ensure effective implementation of social protection programs in the drylands. Incentives are needed to attract and retain qualified staff in hardship posts. Investments in transportation systems and information technology are needed to improve mobility and reduce transactions costs associated with implementing social protection programs in remote dryland areas.

Abbreviations

CFAF	Franc of the African Financial Community
DFID	Department for International Development (of the UK Government)
ECHO	European Commission Humanitarian Aid and Civil Protection Office
FAO	Food and Agriculture Organization of the United Nations
GDP	Gross Domestic Product
HEA	Household Economy Analysis
ICT	Information and Communications Technology
ILO	International Labour Organization
UCT	Unconditional Cash Transfer
UNICEF	United Nations International Children's Emergency Fund
UNOCHA	United Nations Office for the Coordination of Humanitarian Affairs
USAID	United States Agency for International Development

CHAPTER 1

Introduction—The Case for Social Protection in Africa

Following a decade of strong economic growth, stubbornly high rates of poverty have been falling in many countries across Africa. Despite the positive trends, large numbers of people across the continent continue to be affected by external shocks, including droughts and floods. Ethiopia is once again in the news with an El Niño-induced drought affecting upwards of 18 million people.

The vulnerability to shocks is particularly high in the drylands of Sub-Saharan Africa, where poverty rates tend to be higher and livelihoods more dependent on the natural environment. The vulnerability of these areas is explored in detail in the study *Confronting Drought in Africa's Drylands: Opportunities for Enhancing Resilience* (Cervigni and Morris 2016) and in a series of books that complement the study.

This book, one of several prepared as background pieces for the larger Africa Drylands Study, explores the role of social protection in promoting the well-being and prosperity of the populations residing in the drylands of Africa, with a specific focus on the Sahel and the Horn of Africa. It argues that social protection policies and programs, which aim directly to reduce chronic poverty and vulnerability to shocks, have an important role in promoting the resilience of the people residing in these areas. The book reviews recent experience that shows how social protection programs, when well designed and carefully implemented at scale, have the potential to reduce sensitivity and promote coping capacity. Given limitations in scale and scope, however, many countries continue to respond to chronic poverty and predictable needs through the humanitarian appeal system.

The book is structured as follows. Chapter 2 provides an analysis of the poverty and vulnerability context in the drylands and discusses the need for social protection. Chapter 3 outlines the conceptual framework for the book—drawing on the framework used for the Africa Drylands Study—and stresses the important role that social protection policies and programs can play in the drylands in reducing vulnerability and improving the capacity to cope. Chapter 4 analyzes existing social protection programs in the drylands, and chapter 5 assesses the capability of social protection programs to promote opportunities for reducing sensitivity and improving coping capacity. Chapter 6 looks toward building adaptive social protection systems that respond to the needs of people in the drylands. Chapter 7 considers the financing of a national adaptive social protection system and explores the elements of a national social protection system that takes into account adaptation to climate change, including the cost of establishing such a system. Chapter 8 summarizes the main conclusions and recommendations.

CHAPTER 2

Poverty, Vulnerability, and the Need for Social Protection in Africa's Drylands

In recent years, even as many countries in Africa have recorded impressive economic growth and declining rates of poverty, the livelihoods of millions of Africans have remained extremely vulnerable to external shocks. This is particularly the case in dryland areas, where recurring episodes of drought, disease, price spikes, and civil conflict continue to cause widespread food insecurity and hunger. Despite a decade of economic growth, many people in dryland regions of Sub-Saharan Africa remain chronically vulnerable to shocks and poverty.

The breadth and depth of this vulnerability are strikingly evident across the vast region that stretches from the Horn of Africa in the east to the Sahel in the west. In 2011, drought affected nearly 10 million people in Somalia, Kenya, Ethiopia, and Djibouti. In 2013, despite a good harvest, the humanitarian situation in the Sahel remained critical following another season of below average rainfall and high food prices: over 10 million people were food insecure, of whom 1.4 million were children under five affected by acute malnutrition. Today, in 2016, Ethiopia is once again in the headlines as the government appeals for humanitarian assistance for 10.2 million people, while another 8 million receive support through the Productive Safety Net Program. Figure 2.1 shows the number of natural disasters that have hit the Sahel since 1980, and table 2.1 estimates the population affected by each type of disaster during that period.

The challenges faced by those living in the drylands could intensify in the years ahead. Sub-Saharan Africa is one of the regions most exposed to and adversely affected by climate change. Over the past century, the average temperature of the continent has increased by 0.7°C and it is likely to continue increasing by 0.2°C to 0.5°C every 10 years. The impacts of global warming are strongest in the more arid areas of Sub-Saharan Africa and combined with unsustainable soil and water management and limited investment in land protection, climate change is contributing to the further degradation and desertification of drylands (United Nations 2011). Desertification was among the reasons why the 1968–73 drought in the Sahel turned into a devastating humanitarian crisis, and was an important driver of the 2012 food crisis. Climate change hence poses a serious threat to sustained development. It will contribute to lower agricultural

Figure 2.1 Natural Disasters in the Sahel, 1980–2013

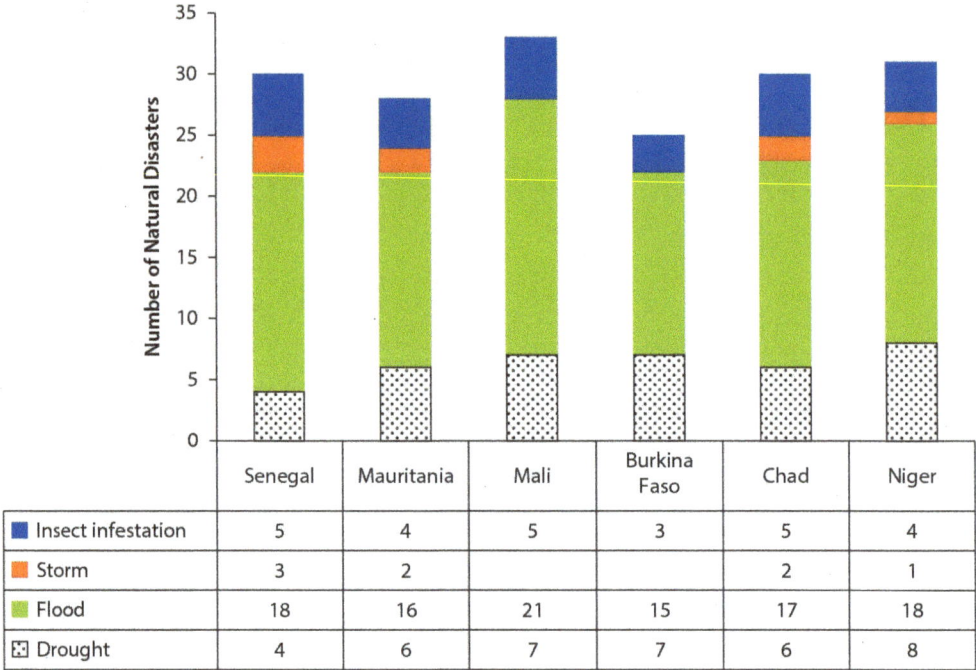

	Senegal	Mauritania	Mali	Burkina Faso	Chad	Niger
■ Insect infestation	5	4	5	3	5	4
■ Storm	3	2			2	1
■ Flood	18	16	21	15	17	18
▣ Drought	4	6	7	7	6	8

Source: Based on data from: EM-DAT: The OFDA/CRED International Disaster Database, www.emdat.be, Université Catholique de Louvain, Brussels.

Table 2.1 People Affected by Droughts and Floods, 1980–2013

	Droughts (millions)	*Floods (millions)*
Senegal	3.3	1.2
Mauritania	4.6	0.2
Mali	6.9	0.3
Burkina Faso	7.0	0.6
Chad	7.0	1.4
Niger	23.6	1.4

Source: Based on data from: EM-DAT: The OFDA/CRED International Disaster Database, www.emdat.be, Université Catholique de Louvain, Brussels.
Note: The disasters accounted for have affected 100 people or more and/or have killed 10 or more. They have resulted in a state of emergency and/or a call for international assistance. Affected people include the injured, homeless, displaced, and/or those requiring immediate assistance during an emergency.

yields and will affect agriculture through different channels, including higher temperatures, increased demand for irrigating water, higher rainfall volatility, and more frequent climate shocks (World Bank 2010). Within these general trends, some groups are particularly sensitive to the negative effects of climate change, as described in box 2.1 and in the sections below.

Box 2.1 Groups Most Sensitive to the Impacts of Climate Change

Vulnerability relates not only to being exposed to the risk of a shock, but also to the degree of sensitivity to that shock. Sensitivity varies with livelihoods and geographical location, but it also depends on the position of power and the ability that a household has to access and use the support it needs. Even when they live in a common area, some households or communities suffer disproportionately more than others from the effects of climate change. They include poor small landholders and the landless, and people whose livelihoods are highly exposed to the impacts of climate-related shocks—such as crop farmers or pastoralists in drylands. They have limited resources or capacities to diversify their sources of revenue, are the first to suffer from food insecurity following weather-related shocks, have little social capital to turn to for support, and are underserved by public or private services (High Level Panel of Experts on Food Security and Nutrition 2012).

Within communities, the impacts of climate change again are not evenly distributed, and some groups are more vulnerable than others. For example, during water shortages, the burden of walking longer distances to fetch water in rural areas or to wait unpredictable hours in water distribution queues in cities often falls onto women and girls, further depriving them from work or schooling opportunities. Similarly, the elderly tend to be particularly affected by heat and malnutrition, especially when there is limited access to healthcare. Finally, children are the most affected by the lack of access to water and by the spread of diseases during droughts, and disasters often lead them to leave school to work and support their families, to get married, or to be left on their own by parents migrating for work (Cabot Venton 2011; Davies et al. 2008).

Aligning with the focus of the Africa Drylands Study, this book concentrates on the subset of people who are vulnerable to poverty, shocks, and the negative impacts of climate change. Drawing on the definitions and methodology laid out in Cervigni and Morris (2016), families and communities that are most *exposed* to the impacts of climate change are defined as those living in dryland areas, that is, classified by the Food and Agriculture Organization (FAO) as hyper-arid, arid, semi-arid, and dry sub-humid. Among them, we consider as "*sensitive to the impacts of climate change and of extreme climate events*" households whose livelihoods directly depend on agriculture, that is, those who are crop farmers, pastoralists, or agro-pastoralists. Finally, we focus on a subset of the sensitive group, defining households *vulnerable* or *unable to cope with shocks* as those with an income below the poverty line of US$1.25 per person per day in an average drought year, below US$1.44 per day during a year of mild drought, US$1.63 during a year of medium drought, and US$1.81 during a year of severe drought.[1]

We also limit the scope of our analysis to the subset of Sub-Saharan African countries with the highest shares of populations living in drylands. All of the focus countries are situated on the southern edge of the Sahara Desert, ranging from

Mauritania on the western coast to Somalia on the eastern tip of the Horn of Africa. Similar to the approach used in the Africa Drylands Study, they are classified into two groups: West Africa and East Africa. Calculations exclude Djibouti, Somalia, and South Sudan, where insufficient data were available for the analysis.

To estimate the share of the population living in drylands that is exposed to droughts, sensitive to droughts, and unable to cope with droughts, and to ensure consistency of analytical approaches among the various components of the larger study, an "umbrella model" was developed as part of the Africa Drylands Study[2] (for details, see Carfagna, Cervigni, and Fallavier 2016). The umbrella model was used to generate detailed projections through 2030 for various scenarios that assumed different levels of economic growth and population growth. The projections discussed in this book correspond to the scenario that assumed medium rates of gross domestic product (GDP) growth and medium rates of population growth through 2030—with both growth rates calculated individually by country.

The proportion of the population exposed to shocks—that is, the proportion residing in drylands—varies between the Sahel and Horn of Africa. Almost 98 percent of the Sahel population lives in arid drylands with a low and declining carrying capacity, and the majority of their livelihoods are weather-dependent. In 2010, a very large share was already living on degraded land, including 24 percent of the population in Niger, 45 percent in Chad, 60 percent in Mali, and 73 percent in Burkina Faso.[3] In the Horn of Africa, 20 percent of Uganda's population lives in drylands, as does about 46 percent in Ethiopia and Kenya, and 63 percent in Tanzania. Figure 2.2 provides a breakdown of the distribution of the populations most exposed to the effects of climate change for 2010 (a total of 166.2 million people), and projected by 2030 under a medium fertility scenario (291.9 million people). It shows that the population in the drylands is expected to decrease in the Horn of Africa and remain the same in the Sahel.

The distribution of those sensitive to climate shocks will depend on differences of livelihoods across the drylands and on the projected evolution of these livelihoods in the face of economic growth.[4] As shown in figure 2.3, the population sensitive to shocks in the Sahel and the Horn of Africa is expected to increase in 2030 under medium estimates of GDP and population growth. In the Sahel, the population is predominantly rural (69.9 percent), with livelihoods based on subsistence farming and pastoralism. Pastoralism remains a predominant way of life in the drylands of the Horn of Africa, although many diverse types of livelihoods also exist there. Larger parts of the populations of Niger and Mauritania subsist on nomadic pastoralism. Differences in livelihoods partly explain the different impacts of shocks: in many areas, such as in Mali's northern region of Kidal, natural and environmental shocks are the main sources of vulnerability, while economic shocks (such as price hikes on markets) are the most cited in the capital Bamako and among households in the wage-earning sector. In Niger, in 2008, more than 50 percent of urban households ranked food price increases as the greatest shock they suffered, whereas the rural fraction was 38 percent. In 2010 the situation reversed and more rural than urban households

Figure 2.2 Population Living in Drylands, East and West Africa, 2010–30 (Million People)

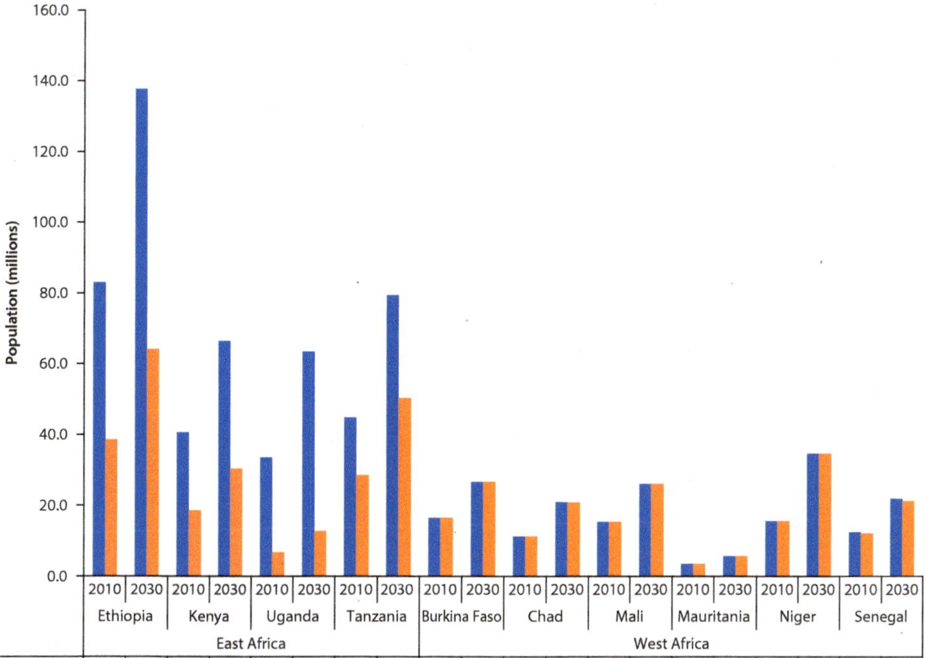

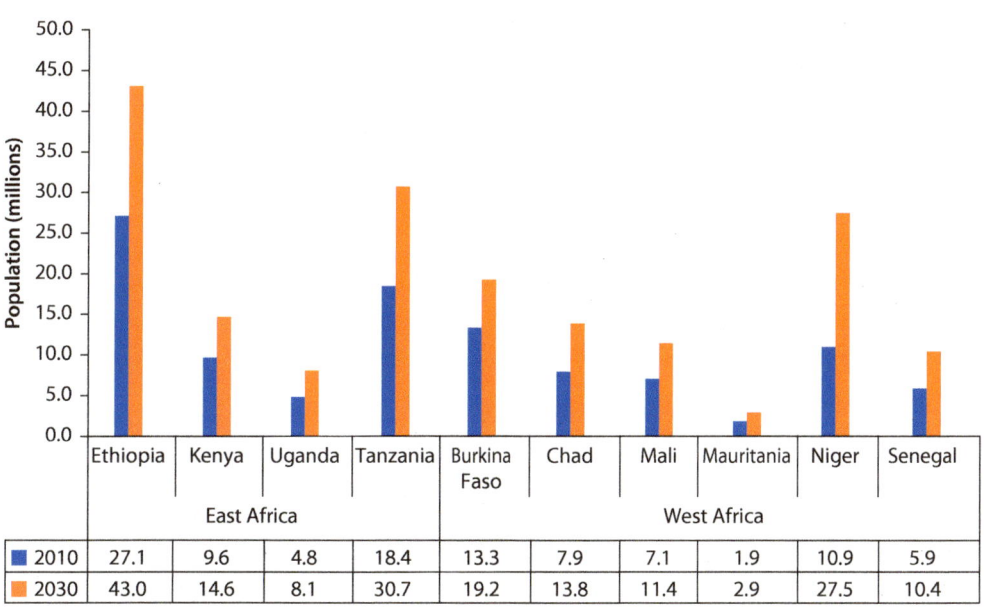

	2010	2030	2010	2030	2010	2030	2010	2030	2010	2030	2010	2030	2010	2030	2010	2030	2010	2030	2010	2030
	Ethiopia		Kenya		Uganda		Tanzania		Burkina Faso		Chad		Mali		Mauritania		Niger		Senegal	
	East Africa								West Africa											
■ Total Pop	83.0	137.7	40.6	66.3	33.5	63.4	44.9	79.4	16.4	26.6	11.2	20.9	15.4	26.0	3.5	5.6	15.5	34.5	12.4	21.9
■ Pop in drylands	38.6	64.1	18.5	30.2	6.7	12.7	28.4	50.3	16.4	26.6	11.2	20.8	15.3	26.0	3.5	5.6	15.5	34.5	12.0	21.2

Source: Carfagna, Cervigni, and Fallavier 2016.

Figure 2.3 Rural Population Sensitive to Shocks, East and West Africa, 2010–30

	Ethiopia	Kenya	Uganda	Tanzania	Burkina Faso	Chad	Mali	Mauritania	Niger	Senegal
	East Africa				West Africa					
■ 2010	27.1	9.6	4.8	18.4	13.3	7.9	7.1	1.9	10.9	5.9
■ 2030	43.0	14.6	8.1	30.7	19.2	13.8	11.4	2.9	27.5	10.4

Source: Carfagna, Cervigni, and Fallavier 2016.

(39 versus 27 percent) said that the greatest shock was price increase (Comité du Pilotage de la Stratégie Nationale de Protection Sociale 2012; World Bank 2004, 2006, 2009a, 2009b, 2011a, 2011c, 2013a). Such variations are also reflected in poverty rates, with poverty highest in arid zones of Niger, where livelihoods depend largely on nomadic pastoralism (see the section below).

Within this context, estimates suggest that large numbers of agriculture-dependent households in the drylands will remain poor—and hence vulnerable to shocks—even in the face of high rates of economic growth, as highlighted in table 2.2 (see also tables A.1 and table A.2 in appendix A, which provides details on levels of vulnerability under different drought severity assumptions). While poverty rates vary from country to country, in the Horn of Africa they are markedly higher among rural populations residing in the drylands, as compared with the national population. In Uganda and Kenya, rates of poverty in the arid and semi-arid rural areas stood at 75 percent in 2009/10 and up to 83 percent in 2005/06, as compared to national averages of 25 percent and 46 percent, respectively. In similar fashion, in many countries, the rates of poverty tend to increase with the Aridity Index.[5]

Table 2.2 Projected Evolution of Vulnerability among Agriculture-Dependent Populations, 2010–30

Population living under US$1.25 per person per day (in million people)	Baseline 2010	2030 low GDP growth	2030 average GDP growth	2030 high GDP growth
East Africa	25.18	42.39	31.81	22.85
Ethiopia	9.96	18.73	12.04	6.80
Kenya	3.72	5.19	4.50	4.13
Uganda	1.79	2.70	2.00	1.27
Tanzania	9.71	15.78	13.27	10.65
West Africa	42.22	86.89	69.53	55.42
Benin	1.07	1.03	0.80	0.49
Burkina Faso	5.53	6.61	5.55	4.46
Chad	2.80	8.03	3.99	3.03
Côte d'Ivoire	0.82	1.26	1.25	1.05
Gambia	0.37	0.77	0.55	0.63
Ghana	0.84	0.99	0.46	0.08
Guinea	0.17	0.26	0.22	0.23
Guinea-Bissau	0.02	0.03	0.04	0.03
Mali	3.57	6.10	5.48	4.85
Mauritania	0.45	0.77	0.60	0.43
Niger	4.41	16.96	15.18	13.65
Nigeria	19.12	37.98	29.90	21.56
Senegal	1.95	3.90	3.51	3.12
Togo	1.09	2.20	2.00	1.81
Grand Total	67.40	129.27	101.34	78.27

Source: Umbrella modeling estimates for East and West Africa (Carfagna, Cervigni, and Fallavier 2016). Countries excluded for lack of data: Djibouti, Eritrea, Somalia, South Sudan, and Sudan. Liberia and Sierra Leone have no drylands.

Figure 2.4 Vulnerability to Shocks in an Average Year, 2010–30 (Million People)

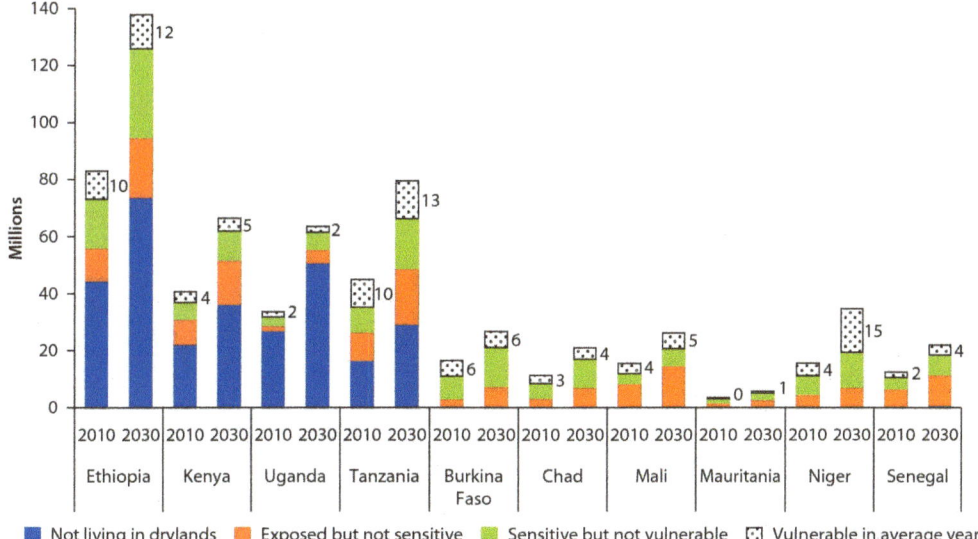

Source: Calculations based on Carfagna, Cervigni, and Fallavier (2016) under scenarios of medium GDP growth and medium fertility rates.

Figure 2.4 summarizes the extent of vulnerability to drought in an average year for each country, as a share of its total population. It breaks down population that is exposed versus non-exposed, extracts the sensitive people, and finally presents the number of people vulnerable to shocks.

Notes

1. The numbers of vulnerable people reported here therefore differ from the numbers that would be presented if other vulnerability traits had been used, for example life-cycle-related vulnerabilities, vulnerabilities related to social exclusion, or characteristics of vulnerable groups living in urban areas.

2. The methodology of this modeling tool is presented in a separate book prepared for the Africa Drylands Study (Carfagna, Cervigni, and Fallavier 2016).

3. See http://hdr.undp.org/en/data, data retrieved January 2014.

4. The umbrella model assumes that there is an inverse correlation between the rate of economic growth and the proportion of people engaging in agricultural activities; that is, countries with higher economic growth will proportionally see the share of their agricultural population (sensitive to shocks) decline.

5. Notable exceptions to this trend are Ethiopia, where rates of poverty are stable across Aridity Index zones, and Nigeria, where poverty rates appear to fall as the Aridity Index increases (as calculated by Zezza and D'Errico from their analysis of living standards monitoring surveys prepared for Cervigni and Morris 2016).

CHAPTER 3

Conceptual Framework—Social Protection Reduces Vulnerability and Improves Capacity to Cope

Social protection systems, programs, and policies help individuals and societies to build resilience to risks, achieve equity, and avail themselves of opportunities. This is done by providing basic income support to the poor and minimizing the risk of households suffering from adverse events while enhancing opportunities for a better future. A range of social protection instruments is used by governments to achieve this objective, as described in box 3.1, with an increasing move among governments and international agencies to look beyond these individual instruments toward supporting the establishment and gradual expansion of social protection systems.[1]

The concept of social protection thus responds directly to the risks and vulnerabilities that arise from the negative effects of shocks—concerns that are at the very heart of the Africa Drylands Study (Cervigni and Morris 2016). Indeed, governments and the international community have increasingly drawn upon

---

Box 3.1 Defining Social Protection Instruments

Social safety nets are non-contributory transfer programs targeted to the poor and vulnerable. These include such programs as cash transfers, public works, and in-kind support like fee waivers and school feeding.

Pension systems provide a minimum income level during old age. They include contributory or non-contributory programs (the latter are often called "social pensions").

Insurance is designed to protect the well-being of households and businesses in the face of adverse events. The types most relevant to social protection in the African context include programs that aim to improve the access of the poor to insurance products, such as those for health and agriculture.

Labor programs and policies promote employment and productivity, particularly among Africa's youth, in the formal and informal sectors. These include, among others, initiatives to enhance the skills of the workforce and to support entrepreneurship and self-employment.

Source: World Bank 2012a.

---

social protection instruments to respond to disasters, both slow onset events, such as drought, and sudden crises, such as earthquakes. The experience from these initiatives has demonstrated that social protection systems that are in place before shocks hit are able to respond in a more timely and effective manner than are initiatives that need to be launched in response to the shock.

At the same time, social protection policies and programs help households and communities build their resilience before shocks hit (see box 3.2 for a general definition of resilience as the ability to overcome shocks). This is achieved at the household level for instance through the regular distribution of cash transfers accompanied by training activities to help diversify livelihoods away from climate-dependent activities, and at the community level through public works that invest in the natural environment—with re-greening initiatives, or in disaster-resilient productive infrastructure, such as dikes or improved irrigation systems. These households and communities weather the negative effects of shocks while helping adapt local livelihoods to changing climate patterns (Kuriakose et al. 2013).

Over the last decade, there has been growing attention for the role that social protection can play in responding to climate change. An obvious first step is recognizing social protection's ability to assist households to reduce their poverty and vulnerability in response to the risks that arise from climate change. In

---

Box 3.2 Resilience as the Ability to Overcome Shocks

Resilience is the ability of a household, community, or country to anticipate, adapt to, and recover from the effects of shocks in ways that reduce its vulnerability, protect its assets, contribute to its recovery, and support its economic and social development (Frankenberger et al. 2012).

The level of resilience of a household or a community depends on the options its members have to cope with shocks, which are a function of their human and financial capital, their ability to produce income, and the availability of relevant public services. A situation of resilience can then be characterized by:

- Diversified livelihood strategies and access to markets;
- Access to financial, social, human, physical, and natural capital;
- Access to quality basic social services;
- Access to social protection programs including safety nets, particularly in difficult periods;
- Access to the information and skills needed to adapt to shocks; and
- Local and national institutions able to adapt to changing realities.

Enhancing resilience therefore means improving households' or communities' economic and social stability by addressing their structural vulnerabilities and increasing their access to services while helping them prepare against future crises (Wernerman 2012).

---

the face of climate change, beyond helping vulnerable communities cope with immediate impacts of more frequent and intense natural disasters, there is a need for households to become more resilient to weather shocks, and to live and thrive under changing climatic conditions. Viewing social protection through the lens of climate change focuses attention toward how social protection can contribute to adaptation via livelihood promotion and diversification and promotion of access to new opportunities. This in turn points to how social protection should be an integrated approach that not only protects people from chronic poverty and the impacts of sudden shocks but also promotes the resilience of vulnerable households.

In an effort to move from concepts to practice, various institutions have sought to put in place frameworks to guide the provision of climate-responsive social protection (see for instance Kuriakose et al. 2013). The current discussions also draw attention to how systems of social protection need to better integrate elements of disaster risk reduction and climate change adaptation (CCA; see Fallavier 2013a; Ovadiya and Costella 2013).[2]

In the Sahel, for example, a World Bank-implemented program was launched, with the support of the UK's Department for International Development (DFID), to help develop adaptive social protection systems specifically adapted to dryland areas consisting of a combination of policies and programs to help poor and vulnerable households reduce the impact of climatic change and other shocks, to build household and community resilience, and to foster access to income-earning opportunities. The main components of the systems it will help set up are:

- **Safety net programs** that can be easily scaled up to respond to climate-related and other types of shocks through the use of conditional and unconditional cash transfers for the poorest and those affected by shocks and public work programs, which can also support climate-resilient infrastructure development in vulnerable areas.
- **Complementary activities** such as training on basic skills and livelihood diversification, as well as accompanying measures aimed at promoting health sanitation practices, nutrition, or early childhood development, to strengthen human capital and resilience of the poor.
- **Linkages to early warning and climate information systems** that can be used for targeting and planning purposes for risk reduction in addition to helping design effective emergency response and adaptation programs.
- **Formal and informal insurance or risk financing mechanisms** that may complement and support social protection systems to build long-term resilience.
- **Labor market policies and programs** that facilitate the employment of poor people in productive income-earning and income-generating activities to help raise living standards, diversify livelihoods, and help households manage risks.

- **Targeting mechanisms** that help identify those most vulnerable to natural hazards and climate change-related risks. Building such a system would allow using climate/hazard information to specifically target ex-ante those that are most at risk of being hit by these types of shocks and to quickly scale up a program in case of necessity.
- **Adequate monitoring systems** that ensure good governance and accountability.
- **Impact evaluations** that help generate systematic knowledge and rigorous evidence on effectiveness of innovations and core components of adaptive social protection systems.

An adaptive social protection approach promotes resilience among households and systems. At the household level, an adaptive social protection approach can help vulnerable people become more resilient to chronic stresses and sudden shocks, in part by adjusting their livelihood systems to the changing climatic constraints. At a systemic level, climate-adaptive initiatives of disaster risk reduction can develop the use of early warning systems and of community-based disaster-preparedness, while social protection approaches can strengthen the economic potential of the chronically vulnerable in ways that are disaster-resilient by diversifying their sources of revenue, for instance with cash-for-work activities in which they build infrastructure limiting the effects of floods or droughts and better irrigating crops. The tools (studies, procedures, or systems) developed for the adaptive system of disaster risk reduction/social protection to identify and reach people most vulnerable to disasters could then be used to scale up social protection and/or humanitarian support in case of acute crises. In turn, the main design features of social protection programs able to effectively and efficiently adapt to natural disaster and climate shocks are:

- **The ability to coordinate across institutions** with effective lines of communication, clearly defined roles and responsibilities for all actors, and linkages and information-sharing arrangements with humanitarian actors;
- **A flexibility that allows programs to be rapidly scaled up and adapted to needs** following a major shock with ex-ante contingency plans and funding mechanisms;
- **The capacity to target** households vulnerable to climate-related risks—and not only the chronic poor;
- **Sound governance and accountability mechanisms,** with the effective participation of beneficiary groups, proper communication and feedback mechanisms, clear guidelines and safeguards to reduce fiduciary risk, and a system of monitoring and evaluation that allows to measure impacts and outcomes and to inform operations; and
- **A focus on building the adaptive capacity of households and communities in the long term**–by increasing communities' physical assets and supporting the development of viable livelihoods at the household level.

The section that follows assesses the state of social protection in the drylands of Sub-Saharan Africa with a view of identifying the building blocks and strategies for moving towards the establishment and expansion of adaptive social protection systems.

Notes

1. For a discussion of social protection systems, see the World Bank Social Protection and Labor Strategy, 2012–22 and the World Bank Social Protection Strategy for Africa (World Bank 2012a, 2012b).

2. The main work linking these three evolving communities of practice into adaptive social protection originates from research conducted by the Institute for Development Study (IDS) and the Overseas Development Institute (ODI) along with DFID and the World Bank since 2006–07 (notably Béné 2011; 2012; Davies and Leavy 2007; Davies et al. 2008, 2013; Davies, Oswald, and Mitchell 2009; Newsham, Davies, and Béné 2011).

CHAPTER 4

Social Protection Programming in the Drylands

Social protection programs are increasingly accepted among policy makers as a key component of poverty reduction strategies. Globally, momentum is growing towards the establishment and gradual expansion of social protection systems. Investments in social protection programs across the drylands of Africa, however, lag behind those made in other developing regions and as a result, their coverage is limited. In the Horn of Africa and the Sahel, most social protection programs are small, fragmented, and largely donor-driven. Still, countries such as Ethiopia, Kenya, and more recently Uganda have scaled up their investments in social protection with encouraging results, providing a model for how other countries can progressively expand coverage to poor, vulnerable populations. At the same time, countries in West Africa are exploring how to establish social protection systems that are designed to adapt to climate change and flexibly respond to the needs of populations vulnerable to shocks.

Coverage of Social Protection

Social Security

The oldest form of social protection programs among countries in the Horn of Africa and the Sahel are national social security schemes for formal sector workers. These schemes are most commonly pensions for civil servants and employees in the formal, private sector. Despite their long history, these schemes cover only a fraction of the population and often tend to benefit better-off segments of the population. In the Sahel, the highest proportion of the population covered by formal social security is at best 13 percent (as reported by governments in United States Social Security Administration 2013). In Uganda, 5 percent of the working age population or 29 percent of wage earners contribute to or are part of a pension scheme (Uganda Ministry of Gender, Labour and Social Development 2014). The pension schemes in Ethiopia cover around 4 percent of the working age population.

More importantly, these schemes generally do not provide effective protection from poverty in old age or due to adverse lifecycle events. In the Sahel, these schemes are under-resourced (in skilled staff and funding). In Kenya, retirement benefits paid by the National Social Security Fund are often deemed to be inadequate (Ministry of State for Planning National Development and Vision 2030 2012), a finding that is echoed across other countries in Africa (World Bank 2012a).

Notably, however, these schemes often constitute a significant share of public spending on social protection. In Kenya for instance, expenditure on civil service pension in 2010 represented about 1 percent of GDP and 88 percent of total government spending on social protection. In Uganda, projections show that government expenditure on the public service pension scheme is likely to more than treble in the long run to 1.1 percent of GDP (World Bank PROST Model).[1] Most Sub-Saharan African countries spend an average of close to 1 percent of GDP on civil service pensions.

Safety Nets

Safety net programs that aim to directly reduce poverty and vulnerability emerged in the Horn of Africa and the Sahel around 2005. Unlike social security schemes, these initiatives aimed to respond to existing high levels of chronic poverty and vulnerability rather than insuring future income streams against the loss of employment in old age or as a result of adverse lifecycle events (where this support exists). In many countries, safety nets were introduced as an alternative to the annual emergency interventions of food aid distribution. There has been a proliferation of safety nets since then, with many continuing to be emergency, short-term responses to acute humanitarian needs. Despite these reforms, the use of food transfers remains common in many countries. In South Sudan, for example, 98 percent of beneficiaries receive safety net support in the form of food,[2] while in Mali targeted food distribution, nutrition, and school feeding programs represented 70–90 percent of safety net spending from 2006 to 2009 (World Bank 2011b).

Even in countries with relatively well-established safety net programs, coverage is low in relation to the size of the population needing support. As shown in table 4.1, by 2014, only three safety net programs in the Horn of Africa could be classified as national in scale, with Ethiopia's Productive Safety Net Program the largest by far, although it reaches less than 7 percent of the population—or roughly 24 percent of the poor; in Kenya cash transfer programs provided support to 15 percent of the absolute poor population in 2014 (assuming perfect targeting of the programs).[3]

The coverage of safety net programs in the Sahel is low and more limited than in the Horn of Africa. There are a few notable programs that reach a large percentage of the total population, but recipients of these programs may not be the ones most in need of support and as a result, large shares of resources are inefficiently used. This is the case with the large-scale distribution of free food or the sale of food at subsidized prices, which are used widely in the Sahel. In Burkina Faso, for example, although the potential number of safety net beneficiaries

Table 4.1 National/Regional Social Transfers Programs in the Horn of Africa (as of 2014)

Program (year established)	Number of people covered	Targeted vulnerabilities	Amount and frequency of payments
Productive Safety Net Program (2010), Ethiopia	7.6 million (in 2009), though scalable up to 8.3 million	Chronically food insecure households who have faced a food gap 3 months or more for 3 consecutive years Community-based targeting	3 kg grain or 17ETB (90 cents) per household member per day (up to 5 days per person a month), six payments (over 6 month period)
Hunger Safety Net Program (2008), Kenya	420,000 (proposed to increase to 720,000 in Phase II)	Chronically food insecure households. Households with more dependents prioritized Community-based targeting 2/3 of recipients are women	4,600KSH ($52), six payments per year (with payments made every 2 months)
Social Assistance Grants for Empowerment (2010), Uganda	600,000 people in 95,000 households in 14 districts	Two transfer types: -Senior Citizens Grant to all elderly 65 and older (60 and older in pastoral Karamoja region) -Vulnerable Families Grant targeted to households with low labor capacity and high dependency ratio	23,000UGX ($9)

Source: Adapted from Lind and Kohnstamm 2014.

could be over 3.9 million people or 60 percent of the poor population, in reality the share of poor people covered by these programs is much lower and the benefits distributed are often too modest to help the poor smooth their consumption (see table 4.2). In Senegal, the effective coverage may be even lower, as 80 percent of the 4 million people who receive some type of safety net assistance obtain it through the national food aid system that distributes free food without proof of need.[4]

Across the Horn of Africa, there are numerous examples of efforts to consolidate and expand safety net coverage. In Ethiopia, the Productive Safety Net Program is expanding to all rural *woredas* in eight of ten regional states, and is targeted to cover 10 million people within the coming three years. The coverage of the National Safety Net Program in Kenya (which brings together the five main cash transfer programs) is expanding. In Djibouti, the United Nations International Children's Emergency Fund (UNICEF) conditional cash transfer program is being taken on by the Djibouti government, itself developing a national program of targeted social safety nets with the support of the World Bank. Even in Somalia, despite the lack of a functioning central government, nongovernmental organizations (NGOs) want to scale up temporal, responsive programs to be long-term and predictable.

Table 4.2 Budget Per Safety Net and Maximum Theoretical Coverage in Burkina Faso, 2009

	Beneficiaries	Budget (CFAF)	CFAF per beneficiary	% of poverty line
Cash and near-cash transfers	359,062	9,021 m	25,124	27
Targeted subsidized food sales	1,800,000	2,144 m	1,191	1
Targeted food distributions	95,878	497 m	5,184	6
Nutrition	377,362	13,771 m	36,493	39
School feeding	1,070,649	8,635 m	8,065	9
Public works	63,065	772 m	12,241	13
Fee waivers for health	140,200	1,249 m	8,909	9
TOTAL (potential coverage)	3,906,216	36,089 m	9,239	10
Share of population covered	25%	Not applicable	Not applicable	Not applicable
TOTAL excluding subsidized sales	2,106,216	33,945 m	16,117	17
Share of population covered	13%			
Share of population below poverty line	43%			

Source: World Bank 2011a.
Note: The poverty line used in 2009 was CFAF 93,949 per person per year. CFAF = Franc of the African Financial Community.

In the Sahel, most safety net programs remain on a pilot basis or are essentially temporary post-crisis interventions focusing on specific geographical areas. Most suffer from weak targeting and inadequate financing that dilute their effectiveness. Overall, there is a strong need for clear frameworks of reflection and action that bring together all the different programs, help select and concentrate on the most efficient, and make full use of a multi-sectoral approach to social protection, in which the actors in charge of delivering basic social services understand the importance of addressing the needs of the most vulnerable, and work with aid agencies to meet these objective, technically and financially. We note, however, that since 2012 the Bank has been launching social safety net programs that aim to become permanent, first in Niger, and from 2014 in Mali. In 2015, similar initiatives will be launched or scaled up in all four other Sahelian nations.

Insurance and Labor Programs

Beyond cash transfer programs, few social protection interventions have been used in the drylands of Sub-Saharan Africa. There are a few small pilots that are looking to experiment with insurance products (particularly those that are index-based) and a very limited number of labor programs, particularly targeting young people. Despite this fact, interest is growing in the possible use of insurance to mitigate against the negative effects of shocks in this area. The later sections of this chapter return to this issue.

Traditional Safety Nets

Traditional safety nets, which have played an important role in mitigating the negative impacts of shocks, are becoming overstretched. Traditionally, taking care of orphans, the elderly, the destitute, or victims of a sudden shock was the responsibility of villages and families, who were to provide them with food, shelters, and survival needs. However, these informal coping mechanisms themselves have largely been stretched to their limits in communities hit by covariate shocks (droughts, price hikes, etc.) that weakened everyone's abilities to support themselves and one another. Even when such mechanisms are still active and do provide essential support—as in the recent case of Mali, where villages in the southern part of the country hosted many internally displaced families—it is at the expense of already very poor host communities. It weakens them while often not allowing beneficiaries to recover to their pre-shock levels of well-being (World Bank 2015.

Spending on Safety Nets

In this context, some countries (such as Ethiopia, Kenya, Uganda, and most recently Djibouti) have made a strong push to establish national safety net programs. In the Sahel, it was not until 2010 that these initiatives were seen as an approach that could be used on a large scale. Despite this move to set up national programs, spending on safety net programs in the Horn of Africa and the Sahel is generally low, even in comparison to other countries in Africa (figure 4.1). Within this general trend, enormous variations exist across countries, reflecting the different scales of coverage, varying payment levels for beneficiaries, different payment modalities (for example, cash and food), and administrative costs. In Burkina Faso, for example, spending on safety nets (excluding subsidies) increased from 0.3 percent of GDP in 2005 to 0.9 percent in 2009 (World Bank 2011a).

In all countries, a significant proportion of funding for safety net programs is from donor agencies. In Niger, donor funding accounts for 70 percent of all expenditure on safety nets. This reflects a broader trend in Africa, although there has been a recent move in Kenya (with a similar commitment in Ethiopia) to

Figure 4.1 Government and Donor Spending on Social Safety Nets as a Percent of GDP, Selected Countries

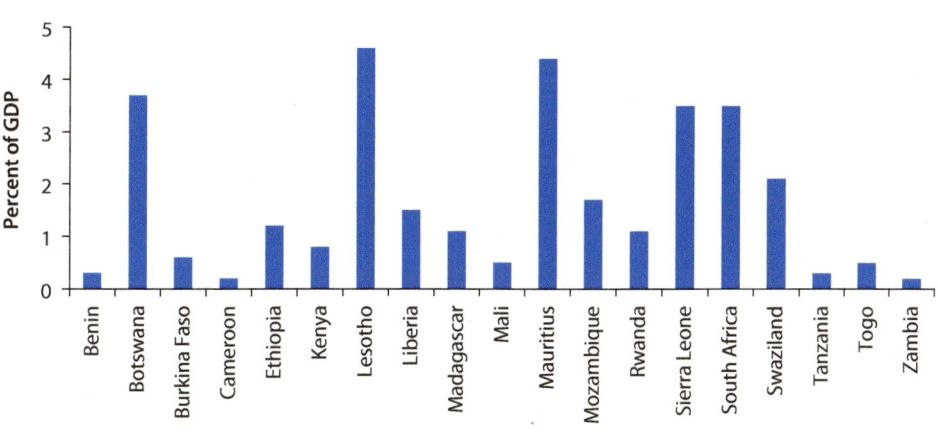

Source: Monchuk 2014.

Table 4.3 Cost and Financing of Safety Nets in the Sahel (2008–12)

	Spending on safety nets (% of GDP)		Share of government spending (excl. subsidies)	Share financed by		Notes
	Excluding subsidies (%)	Including subsidies (%)		Government (excl subs.) (%)	Donors (%)	Years
Burkina Faso	0.6	1.3	<1.0%	20	80	Ave. 2005–09
Mali	0.5	0.5		40	60	Ave. 2006–09
Mauritania	1.3	3.2	4.6%	62	38	Ave. 2008–13
Niger	Not available	Not available	1.0–5.0%	33	67	Ave. 2001–06
Senegal	0.27	Not available	Not available	34	62	Ave. 2009–11

Sources: Prepared by Annamaria Milazzo, from country safety net assessments. Figures include donor financing except general budget support. Excludes funding by the private sector.

increase government financing to safety net programs. Table 4.3 presents the relative contribution of governments and donors to safety nets among countries in the Sahel.

Capacity of National Social Protection Programs to Respond to Dryland Vulnerabilities

Since safety net programs in the Horn of Africa and the Sahel have historically been used to respond to chronic poverty, the coverage of safety nets is often higher in dryland areas than in other areas. For example in Kenya, because the government has tried to prioritize poorer areas when extending safety net support, coverage rates among absolute poor households in the four arid- and semi-arid counties in northern Kenya exceed 40 percent.[5] In Uganda, much of the safety net programming (which is highly fragmented) focuses on the north, which includes the semi-arid areas of Karamoja.[6]

In response to the inefficiencies of earlier safety net systems, more recently a number of countries have begun experimenting with new models. In 2011, Niger began to provide chronically poor households with regular cash transfers over an 18–24 month period, with the goal of helping them meet basic consumption needs while gradually building their human capital. Over time, similar programs were introduced in other Sahelian countries. The model is simple—cash is provided along with accompanying measures, such as education to raise nutritional awareness among mothers, or training to instill employable skills among working-age youth and adults. The programs are designed to be flexible, so that the amounts of the transfers and the types of accompanying measures can be adapted to local needs, and so that coverage can be scaled up in times of crisis. While this new generation of safety net programs shows signs of promise, most of the programs are still at early stages of implementation, and are not yet ready to be scaled up rapidly in response to a crisis. The experience in Niger was generally positive, although an important lesson that has emerged is that a one-size-fits-all approach is not always effective, as permanent programs and emergency responses need to be adapted to the diversity of livelihoods systems found throughout the country. Such lessons highlight the need for national programs to be tailored to respond to the sources of risk and vulnerabilities in the drylands, as livelihoods patterns can vary significantly across these areas (map 4.1).

Several countries in East Africa have made efforts to tailor safety net programs to meet local needs. In Kenya, for example, the Hunger Safety Net Program was designed specifically to respond to the vulnerabilities of people living in the arid and semi-arid areas of the northern part of the country. The Hunger Safety Net Program uses cell phone-based technology to support a mobile payment system that is adaptable to pastoral livelihoods. In Ethiopia, under the Productive Safety Net Program, efforts have been made to tailor the design and delivery of assistance to the pastoral regions of Afar and Somali. These efforts were launched within the parameters of an existing program, however, and because certain features of the pre-existing program proved inflexible, the results were mixed. Despite the variable results, the Productive Safety Net Program provides a rare example of a safety net program that has attempted to tailor the design and delivery of public works to pastoral livelihoods (Lind and Kohnstamm 2014; World Bank 2010).

Beyond cash transfer programs, few social protection interventions have specifically targeted the drylands. A growing body of evidence suggests that insurance and labor programs can assist households to better mitigate the impact of shocks and diversify their livelihoods, yet such programs have rarely been introduced in dryland areas, and those that exist tend to be small, pilot initiatives (for example, in Ethiopia, Kenya, Senegal, and Nigeria). In Niger, a model of *warrantage* has shown its potential: instead of selling their crops at a low price just after harvest, farmers pool part of their crops in common granaries against which they obtain a group loan from microfinance institutions. These loans are paid later, when the crops are sold when prices are higher (Fallavier 2013b). In Kenya, a pilot with an index-based livestock insurance product was designed to protect against drought (Chantarat et al. 2013). The Rural Resilience Initiative in Ethiopia also provides weather-indexed insurance. In 2012, some 12,200 small-scale farmers received a payment in Tigray. Though the scale of these initiatives is still limited, they will be important for generating evidence to inform future programming (Lind 2013).

Among African countries, only Ethiopia has established the capacity to rapidly expand the coverage of its safety net program in response to shocks. This capacity is critically needed in dryland zones, where large numbers of poor people are often exposed to droughts that can suddenly undermine their livelihood strategies. In Ethiopia, rapid scalability of the Productive Safety Net Program is ensured through contingency funds that are held at district (*woreda*), regional, and federal levels, although the use of these budget lines has evolved over time. These contingency funds can be used by local officials to respond to transitory food insecurity, including food insecurity arising from drought. Beginning in 2008, the

Map 4.1   Comparing the Diversity of Rural Livelihoods Between Senegal and Niger

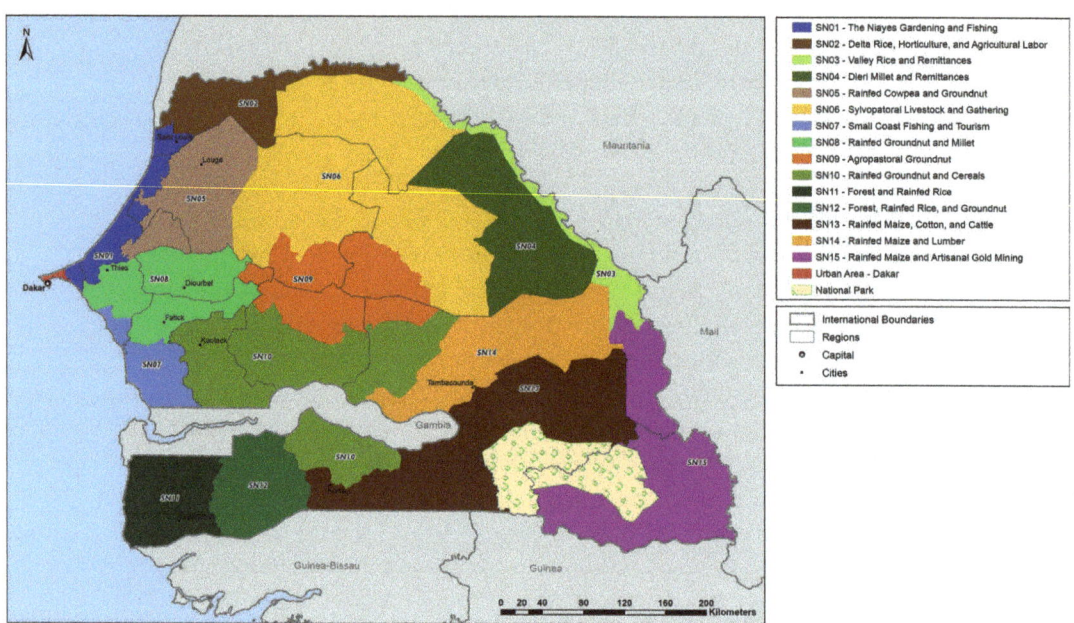

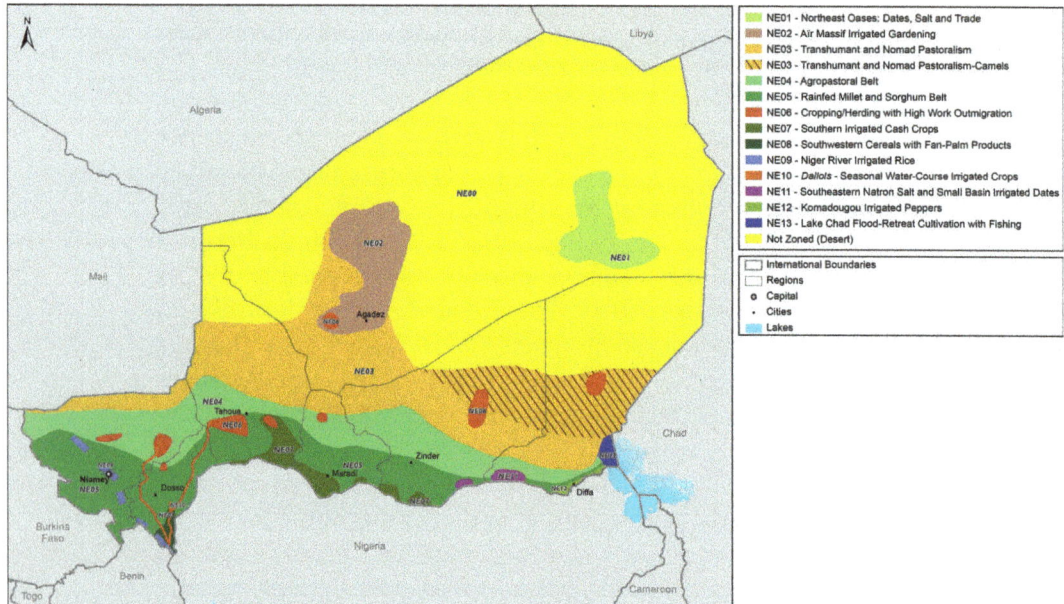

Source: The FEWS NET Project, USAID, 2015 (for Senegal) and 2011 (for Niger). Used with permission; further permission required for reuse.

contingency funds have been complemented with a risk financing mechanism that allows the federal government to trigger the release of additional resources to increase the value or frequency of transfers to existing beneficiaries and to provide support to additional people negatively affected by drought.

The scalability feature of the Productive Safety Net Program was designed to provide a first line of response to drought, complementing the existing humanitarian appeal mechanism, which will continue to be used to respond to needs in areas outside the Productive Safety Net Program districts or in cases where needs within the Productive Safety Net Program districts exceed available resources. During the 2011 Horn of Africa crisis, the administrative and logistical infrastructure of the Productive Safety Net Program proved capable of scaling up the coverage of the program very rapidly, thereby strengthening the capacity of hundreds of thousands of vulnerable households to withstand a series of unexpected shocks.

In 2015, the Government of Ethiopia decided to replace the regional contingency budget and risk financing mechanism with a single, federally managed contingency budget line. In this way, the federal government can flexibly deploy these resources each year to respond to needs arising from drought and other shocks irrespective of where in the country these occur.[7] It is anticipated that these funds will be used each year to respond to transitory food insecurity; need above this level is to be sourced through the humanitarian appeal system. The Hunger Safety Net Program, which operates in four counties in northern Kenya, recently created a mechanism by which it can scale up in response to drought. Through this mechanism, the population of northern Kenya has received cash transfers delivered through the program systems in response to drought.

Humanitarian Response as a Social Protection Instrument

Because social protection programs are generally very small—and because few have the capacity to scale up rapidly in response to shocks—most governments rely on food aid and on humanitarian appeals as a means of responding to chronic poverty and predictable emergencies. Among countries in the Horn of Africa, food aid has been the dominant form of support for poor and vulnerable households in all countries. In Ethiopia, distribution of humanitarian aid was the annual response to chronic food insecurity, costing on average US$265 million per year prior to 2005, when the Productive Safety Net Program was introduced. Similarly, from 2005 to 2010, spending on food aid accounted for 53.2 percent of all government spending on safety nets in Kenya. Humanitarian assistance to the Sahel increased from US$37 million in 2000 to US$630 million in 2010 (figure 4.2). In 2014, assistance reached US$878 million (45 percent of the actual needs of US$1.95 billion), and for 2015 the UN Office for the Coordination of Humanitarian Affairs (UNOCHA) evaluates the humanitarian needs at US$1.96 billion. Of this, about 50 percent would be for food security and nutrition and 20 percent for support to displaced people and refugees (UNOCHA 2015).

Figure 4.2 Humanitarian Aid Received, Selected Countries, Horn of Africa and Sahel (2000–11, USD million)

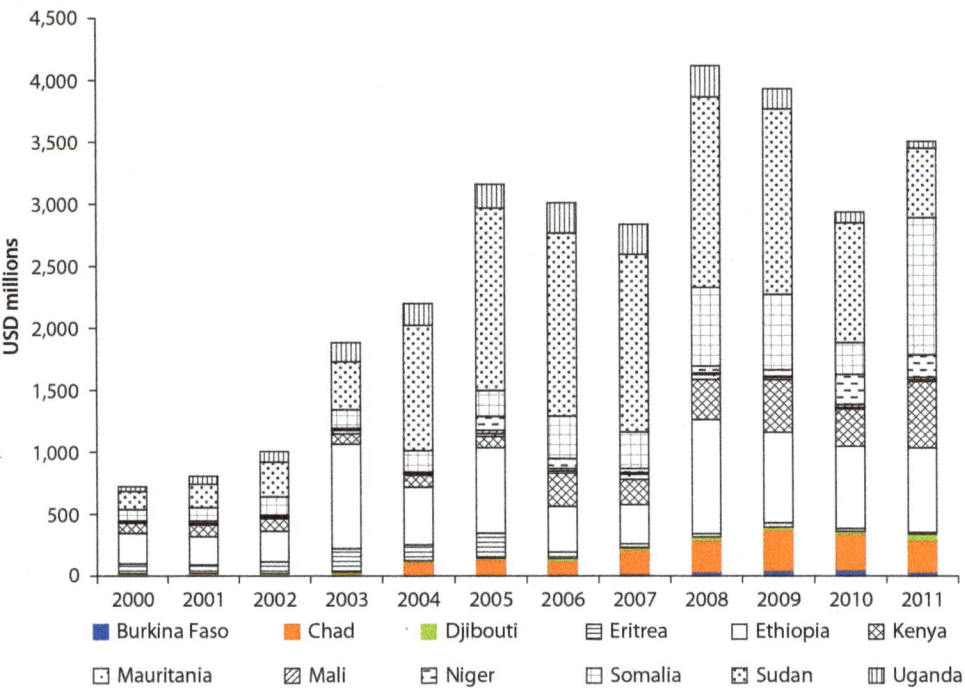

Source: Calculations based on http://www.globalhumanitarianassistance.org/data-guides/datastore.

Humanitarian assistance in the drylands typically involves the provision of food, cash, and other in-kind resources and services to help affected households cope with the immediate effects of drought. Delivery mechanisms for humanitarian aid often consist of food distribution, cash transfer programs, feeding programs, purchase of livestock, and provision of health services and water and sanitation services. Humanitarian assistance is an appropriate short-term response to emergencies, but in many countries it is provided year after year in the same areas and to the same recipients, suggesting it is used as a long-term instrument to address chronic poverty. This use of humanitarian aid is inappropriate as the delivery costs tend to be extremely high. Food aid, for example, is usually procured internationally and transported across long distances, making it very expensive. In Ethiopia, prior to 2005 when the Productive Safety Net Program was introduced, food distribution programs were the annual response to chronic food insecurity, costing on average US$265 million per year. In Kenya, spending on food aid from 2005 to 2010 accounted for 53.2 percent of all government spending on safety nets. Given the high cost of delivering food aid, it is estimated that every dollar spent on food aid could have generated twice as many benefits to recipients had it been provided in the form of a permanent cash transfer program.

In addition to being expensive, protracted use of humanitarian assistance is often ineffective. While emergency distribution of food can save lives, it faces implementation challenges. The food often arrives late and the amount delivered is generally less than what is required. Additionally, given the emergency nature of the support, it is often difficult to target the poorest and most vulnerable households; the authorities tend to focus mainly on getting the resources to communities that have been especially hard hit, but the allocation of resources to households within these communities is often done in an *ad hoc* fashion, or the resources are made available to all households, regardless of need. Finally, because humanitarian aid resources become available only after a shock has occurred and donors have had time to respond to appeals, the timing and amount of transfers received by the affected households tend to be inadequate to meet all of their needs.

Additionally, as shown in table 4.4, in the Sahel, food transfers remain the preferred type of safety net transfer in many countries. Governments use a range of targeted subsidized food sales, free food distributions, nutrition programs, and school feeding programs. The typical safety nets in the Sahel focus largely on food support in response to sudden crises. However, if governments allocate the largest shares of their safety net budgets on programs aiming to make food more readily available, these do not specifically target the poor. They are mainly universal subsidies to food and fuel prices and promotion of universal coverage of school feeding. Figure 4.3 shows how even when not counting universal price subsidies, 75 percent of Burkina Faso's safety net expenses are spent on subsidized food sales and school feedings. Mali (figure 4.4) uses more "targeted" food distributions (with an increase also in subsidized food sales in 2008), but in both cases—as in all other countries—with very limited means of targeting, it is doubtful that the food reaches the people who most need it (World Bank 2011a, 2011b).

Note that these numbers are indicative only. Not all expenses are fully accounted for in each country, and categories may overlap.

Table 4.4 Shares of Social Safety Net Spending by Program Type, 2008–12 (Percent)

	School feeding	Public works programs	Fee waivers	Cash transfers/ vouchers	Food/in-kind distribution	Nutrition programs
Burkina Faso	51	6	2	5	7	29
Mali	18	15	0	0	36	31
Mauritania	12	8	7	5	59	8
Senegal	80	n/r	n/r	7	4	7

Sources: Prepared by Annamaria Milazzo, from country safety net assessments. Senegal: calculation by the author from Senegal safety nets assessment. n/r for "no record."

Figure 4.3 Breakdown of Safety Net Programs by Type in Burkina Faso, 2009

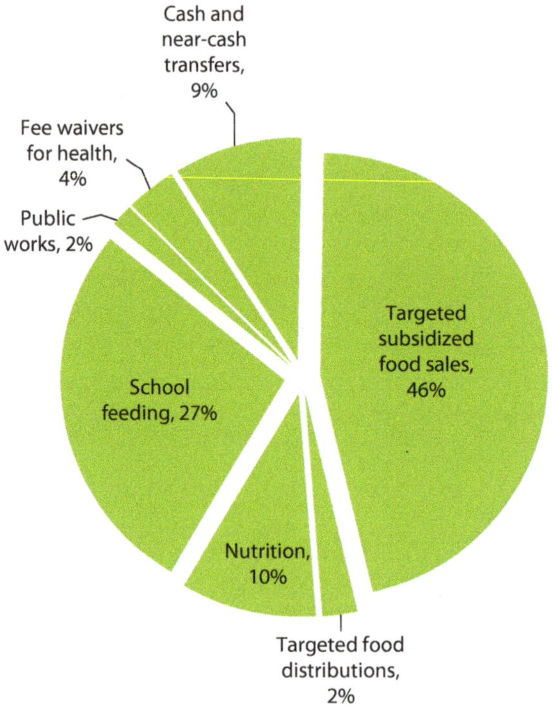

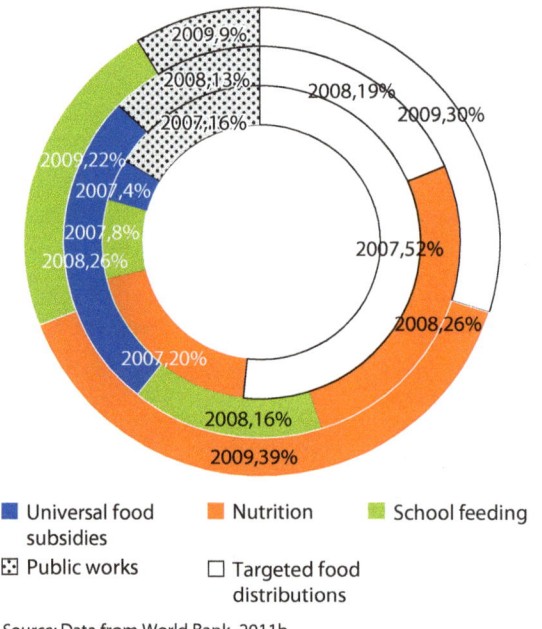

Cash and near-cash transfers, 9%

Fee waivers for health, 4%

Public works, 2%

School feeding, 27%

Targeted subsidized food sales, 46%

Nutrition, 10%

Targeted food distributions, 2%

Source: Data from World Bank (2011b).

Figure 4.4 Evolution of Safety Net Use in Mali, 2007–09

2009,9%

2008,13%

2007,16%

2008,19%

2009,30%

2009,22%

2007,4%

2007,8%

2008,26%

2007,20%

2007,52%

2008,26%

2008,16%

2009,39%

■ Universal food subsidies

■ Nutrition

■ School feeding

⊞ Public works

□ Targeted food distributions

Source: Data from World Bank 2011b.

Note: 2007 is inside ring, 2008 middle ring, 2009 outside ring.

Notes

1. This assumes no change to the system and its governing parameters over the next 50 years. This estimate does not include spending on military pensions, for lack of data.

2. Unless otherwise referenced, the figures on coverage of social safety nets in East Africa result from data collected from service providers (government and aid agencies) for the larger drylands report (Cervigni and Morris 2016) and reported by Lind (2013). Data on Sahelian countries are from World Bank social safety net studies as well as from data provided by the World Food Programme and UNICEF.

3. Calculations carried out to inform the roll-out of the Government of Kenya's National Social Protection Program.

4. The Government of Senegal is the first in the region to have measured and acknowledged the inefficiency of its universal subsidies on fuel and some foods, and it agrees with its aid partners that developing a better system of targeted safety nets will be more efficient in addressing vulnerabilities. It will, however, have to ensure that before transitioning from universal subsidies to cash transfers, the retail and import markets are developed for fuel and imported staple commodities (see more in World Bank 2013a).

5. Unless otherwise referenced, estimates of safety net coverage in East Africa are based on primary data collected expressly for this book from service providers (for example, government agencies and aid agencies). Estimates of safety net coverage in the Sahelian countries come from World Bank social safety net studies, as well as from updated data provided by the World Food Programme and UNICEF.

6. Uganda Ministry of Gender, Labor and Social Development, 2014.

7. In practice, the federal contingency budget of the Ethiopia Productive Safety Net Program is used in the regions of Ethiopia that participate in the program.

CHAPTER 5

Opportunity for Reducing Sensitivity and Improving Coping Capacity

Social protection programs, when correctly designed and effectively implemented, can reduce vulnerability in the drylands by reducing the sensitivity to shocks of vulnerable households and by improving the capacity of these households to cope after a shock has occurred. It is important to distinguish between these two objectives—reducing sensitivity and improving coping capacity—and to consider the characteristics (including the financing requirements) of the different types of social protection programs that can be used to achieve each objective.

Reducing Sensitivity

Social protection programs can reduce sensitivity to shocks by enabling poor and vulnerable households to invest in human capital, build assets, and diversify their livelihood strategies. The social protection programs that perform this function are those that target the chronic poor and provide continuous assistance over a sustained period. Sustained, predictable support provides the certainty households need to enable them to take risks that can lead to higher returns on investments and enhanced income streams. Predictable, multi-annual social protection support to households has also been shown to stimulate investments in human capital and assets that can, over the longer term, lift households out of poverty. While the assistance is provided over multiple years, the expectation is that for individual households it is finite, in the sense that it will be suspended once the household has built an asset base and diversified its livelihood strategy, because at that point the household will be resilient and will no longer require support. These objectives are more effectively achieved when social protection support is combined with investments in human capital and livelihoods, and when it is integrated with other development programs, such as those proposed for dryland areas.[1]

Cash Transfers

One type of social protection program that allows households to invest in human capital, build assets, and diversify their livelihood strategy involves cash transfers. Cash transfers may be unconditional or conditional. Unconditional cash transfers provide greater flexibility for recipients to use the money to address their own priorities, but they bring the risk that the resources may be used for immediate consumption, instead of being invested in ways that would allow the recipients to improve their livelihoods in future years. Conditional cash transfers are designed to overcome this problem by requiring recipients to engage in activities that are likely to generate benefits over the longer term. Increasingly, the delivery of support is complemented with other services, such as those that promote nutrition or provide skills training. This approach is becoming particularly common in the Sahel. These programs, when properly designed, can support more productive and potentially more diversified livelihoods and help beneficiaries participate in the growth process and take advantage of the investments made in agriculture, pastoralist activities, etc., proposed in this report.

A large and growing body of evidence shows that cash transfer programs work, including in dryland areas. In arid and semi-arid zones of northern Kenya, households receiving regular cash transfer support from the Hunger Safety Net Program withstood a severe drought in 2011 without any increase in poverty levels, whereas among those not receiving cash transfer support, 5.3 percent of households had fallen into the bottom income decile following the drought. In Ethiopia, the average period during which households participating in the Productive Safety Net Program reported being food secure increased from 8.4 months in 2006 to 10.1 months in 2012. While it is not possible to disaggregate these findings by aridity zone, data from regions in Ethiopia that are predominately classified as drylands show results that are similar to those recorded in more humid regions (Hoddinott and Lind 2013).

Public Works

A second type of social protection program that can help households reduce their sensitivity to shocks is public works. In addition to delivering immediate assistance to participating households by paying wages, public works can put in place productive infrastructure that can permanently improve the livelihood strategies of recipients. Public works programs are particularly common throughout the Horn of Africa. More than a decade of experience with public works programs in Ethiopia has demonstrated how watershed development schemes have the potential to transform the natural environment and enhance the resilience of communities and households, especially when they are designed using community-based planning approaches and implemented over multiple years. Through public works initiatives, the Productive Safety Net Program has constructed 600,000 kilometers of soil and stone bunds that enhance water retention and reduce soil erosion. Public works initiatives supported under the Productive Safety Net Program have also been used to protect 644,000 hectares of land in area enclosures, leading to improved soil fertility and increased carbon sequestration. Within these enclosures, groundwater levels are rising, springs last longer into the dry season, and woody and herbaceous vegetation have increased. These results are having a direct impact on rural livelihoods (World Bank 2014).

Insurance Programs

A third type of social protection instrument that can reduce sensitivity to shocks is programs that facilitate access to insurance products that lower the risk associated with traditional livelihood strategies, such as farming and livestock keeping. Typically these products are designed to provide protection against extreme weather events, including drought, by linking payouts to weather-based indexes. While these have been tested only on a limited scale through pilot schemes, experience suggests that well-designed weather-indexed insurance programs can be effective in protecting rural households from shocks. In Kenya, for example, when drought triggered payouts by the Index-Based Livestock Insurance, the frequency with which households protected by the scheme engaged in negative coping strategies (such as selling livestock or reducing the number of meals eaten each day) fell by 33 percent, and the frequency with which they engaged in distress sales of livestock fell by 50 percent. A 33 percent drop in food aid reliance was also observed. In Ethiopia, evaluations of households insured through the Rural Resilience Initiative concluded that compared with non-participants, farmers who bought insurance planted more seeds, used more compost, adopted modern varieties at higher rates, used less family labor and more hired labor, diversified their income sources, and experienced smaller losses of livestock (Hoddinott and Lind 2013). If the experience gained through these pilot schemes can be harnessed to build effective large-scale insurance programs, the coping capacity of households living in drylands could be further strengthened. Over time as they become confident that insurance products can provide effective protection against the negative effects of shocks, households will be encouraged to invest in more productive livelihood strategies that will reduce their chances of falling into poverty.

Improving Coping Capacity

In addition to reducing sensitivity to shocks, social protection programs can improve coping capacity and help households recover after a shock has hit by providing immediate assistance, usually in the form of food or money. Unlike other types of social protection programs that target the chronic poor and provide continuous assistance over a sustained period, this second type of program— often referred to as "temporary" safety nets—is designed to provide short-term assistance to help affected households cope with the effects of a specific shock. Unlike other types of programs that are designed to encourage households to invest in human capital, build assets, and diversity their livelihoods, this type of program allows immediate needs to be met by providing consumption support, thereby allowing households to avoid the use of short-run coping strategies that will undermine their livelihoods over the longer term, such as selling livestock or pulling their children out of school.[2] It is important to note that this type of program is not expected to have a permanent effect on the poverty status of beneficiary households, although households may avoid falling deeper into poverty. Therefore, households that receive benefits through this type of program will not necessarily be resilient in subsequent years, after the flow of benefits has stopped.

Because this second type of program is designed to improve coping capacity by taking action when a shock is imminent or after a shock has hit, it is critically important that whatever instruments are used be part of the permanent system and that they be rapidly scalable. In addition, it is important that scalable safety net programs be coordinated with humanitarian support, so that humanitarian support can be mobilized quickly when the capacities of scalable safety net programs are exceeded. The next section explores these types of national systems.

Notes

1. Even so, for some households, depending on the context, this process can take a long time.

2. The costs of not protecting poor populations from the negative effects of shocks are high and long lasting. Ethiopian households that suffered during the 1984/85 drought continued to experience 2–3 percent less annual per capita growth in the 1990s. Children in households in Burkina Faso that experience a negative income shock are less likely than other children to enroll in school. The negative consequences of reducing investments in children can be irreversible: malnutrition alone lowers GDP growth by 2–3 percent.

CHAPTER 6

Building Adaptive Social Protection Systems That Respond to the Needs of People in Drylands

The core of any successful safety net system is a national safety net program that has the ability to scale up coverage rapidly and efficiently. Currently in Africa, safety nets are at different stages of development (table 6.1). In the Horn of Africa, a number of countries (including Ethiopia, Kenya, and Uganda) have made progress in putting in place national safety net programs. While the rationale for these programs and their features differ, each country has established a government-led safety net program that is national in scope. These initiatives can serve as examples to the many Sahelian countries that have yet to introduce

Table 6.1 Country Typology Based on Crisis Preparedness and Safety Net Capacity

	Strong measures to improve SSNs during a crisis	Moderate measures to improve SSNs during a crisis	Limited or no measures to improve SSNs during a crisis
Tier I **No SSNs in place**		Comoros	Central African Republic, Chad, Congo, Dem. Rep., Cote d'Ivoire, Equatorial Guinea, Eritrea, Gambia, Guinea, Mauritania, Somalia, South Sudan, Sudan
Tier II **Weak capacity in SSNs**	Niger, Tanzania, Zimbabwe	Ghana, Liberia, Malawi, Mozambique, Sierra Leone, Togo, Uganda	Angola, Benin, Burkina Faso, Burundi, Cameroon, DRC, Gabon, Guinea Bissau, Madagascar, Mali, Nigeria, São Tomé, Senegal, Swaziland, Zambia
Tier III **Increasing capacity in SSNs**	Ethiopia, Kenya, Rwanda	Cape Verde, Lesotho, Mauritius	
Tier IV **High capacity in SSNs**		Botswana, Namibia, South Africa	

Source: Adapted from Monchuk 2014.
Note: SSN = social safety net.

safety net programs, as well as Somalia, Sudan, and South Sudan, whose investments in safety net programs have been modest.

Since the incidence, severity, and impacts of many shocks cannot be predicted, scalability is of paramount importance in the design of safety nets. To be effective, a national safety net program must be capable of rapidly scaling up the provision of transfers to people who have been (or will be) negatively affected by a shock. The best scalable safety nets are able to quickly respond to an imminent or emerging crisis on the basis of information generated through early warning systems and seasonal assessments.

Scaling up of one or more existing programs allows for a much faster and effective response to drought and other emergencies than is possible using the traditional humanitarian appeal process. Additionally, transfer systems that are already in place can have a greater impact in terms of consumption smoothing and livelihood protection per dollar spent than expensive *ad hoc* programs. Investing in early warning systems is central to this approach, to ensure that a reliable and transparent stream of information is the basis for triggering any response.

To date, however, few safety net programs in Africa have the capability to respond to shocks. Ethiopia's Productive Safety Net Program is at the forefront in this regard, while Kenya is building rapid response capacity into its National Safety Net Program, particularly within the Hunger Safety Net Program. The Ethiopia model currently uses contingency funds to create a pool of resources at district and national levels that can be accessed during times of crisis to increase the coverage of the program or duration of support to beneficiaries. The Kenya model is exploring different mechanisms for securing contingency financing, but will include the government's National Drought Contingency Fund.

The success of these models hinges on a well-functioning early warning system that forecasts a crisis or provides detail on the areas and people most affected and clear, objective "triggers" that determine when resources should be deployed. In Niger, the government has put in place the components of an early warning system against food insecurity and malnutrition and a range of early response mechanisms to minimize the impacts of food crises ahead of time and to be ready with humanitarian response when crises hit.

This mechanism is complemented by the creation of a unified registry of safety net beneficiaries and the contingency planning capacity of the social safety net program.[1] The creation of a unified registry—together with a payment mechanism that covered all households in the program area—enabled the Hunger Safety Net Program to scale up quickly in response to drought. Where such systems do not exist, safety net programs can use existing mechanisms to identify and reach those in need of support. In many cases, humanitarian response programs offer insights and options into how this can be achieved.

More broadly, to strengthen resilience, social protection should be integrated with disaster risk reduction initiatives and emergency relief operations through a common understanding of the different types of vulnerabilities to shocks, better early warning and planning of coordinated early action, and development of a capacity to quickly target and reach people affected by crises. To better factor in disaster risk reduction, be ready to support humanitarian intervention, and adapt to climate change, national social protection systems should: (i) develop early warning information systems that can be used both for climate-related and socio-economic risks and regularly update contingency plans for early intervention in preventing crises; (ii) develop funding mechanisms to scale up activities in time of emergency; and (iii) ensure that permanent social protection initiatives include risk reduction, preparedness, and adaptation to climate change.

In this way, social protection becomes part of a larger integrated system of risk management that links disaster risk reduction with risk-informed development programming and links prevention and development with humanitarian responses. Box 6.1 shows how social protection, disaster risk reduction, and emergency humanitarian aid contribute to complementary objectives. Developing tools that can serve both the goals of risk reduction, social protection, and emergency humanitarian aid for analysis, program planning, and responses can improve their effectiveness. They include common mechanisms to identify and target beneficiaries such as poverty and vulnerability assessments and mapping, livelihoods analyses, and the use of shared registries of beneficiaries. Programming for development and social protection activities in ways that account for the risks of crisis and the use of early warning and early action systems also helps identify, prioritize, and address the risks that could affect vulnerable populations. It further helps in promptly intervening in the right places with adapted instruments during crises, and it provides directions for long-term social protection initiatives to strengthen the resilience of households and communities to shocks and stresses.[2]

Of the Sahelian countries, Niger has been hit the hardest by repeated crises since 2005. In response to this situation, after 2008 it became one of the most active countries in piloting interventions aimed at limiting the impact of crises on the most vulnerable populations. The government developed the components of an early warning system against food insecurity and malnutrition and a range of early response mechanisms to minimize the impacts of food crises ahead of time and to be ready with humanitarian response when crises hit. Its *National Mechanism for the Prevention and Management of Disasters and Food Crises* has shown its value from 2011 and is being gradually strengthened with the development of different mechanisms, such as creation of a unified registry of safety net beneficiaries and the contingency planning capacity of the social safety net program to support humanitarian actors in undertaking rapid responses in some of the most vulnerable areas of the country.

In the same period, several initiatives were developed at the regional level to produce information in preparation and response to crises across Sahel's dryland nations. Some are internal to specific agencies, others shared. UNICEF's *Early Warning Early Response System* allows the agency and its partners to respond to possible emergencies at the regional and national level. FAO maintains a *Global Information and Early Warning System* that provides up-to-date information on food and agriculture production, on markets and prices, as well as on nutrition in countries affected by food insecurity (see http://www.fao.org/giews/). European Commission Humanitarian Aid and Civil Protection Office (ECHO) is developing its own internal Emergency Response Centre to anticipate and react promptly to emergencies and disasters, and is supporting the Sahel Household Economy Approach portal (http://www.hea-sahel.org), which shares HEA tools and studies in the region. United States Agency for International Development's (USAID's) Famine Early Warning System Network (FEWS NET) (http://www.fews.net) provides early warning and vulnerability information on emerging and evolving food security issues, while United Nations Office for the Coordination of Humanitarian Affairs (UNOCHA's) Reliefweb (http://reliefweb.int) makes extensive humanitarian information on crises and disasters available on a real-time basis.

The convergence in the use of instruments (cash transfer programs) and of methods to identify vulnerabilities and target potential beneficiaries as well as the development of early warning systems has laid the ground for the development of disaster risk reduction initiatives, and for improving coordination between risk reduction, long-term social protection, and emergency responses.

Box 6.1 Synergies between Social Protection, Disaster Risk Reduction, and Humanitarian Aid

To integrate efforts in social protection, disaster risk reduction, and humanitarian intervention to build resilience against shocks, it is useful to point out how they contribute to complementary objectives:

- **Social protection builds the resilience of vulnerable families** by enabling their access to basic services and providing them social transfers vital to overcome shocks and invest in their human capital. By helping governments provide basic services and recognize the rights of all, it also contributes to strengthen social cohesion, and to support state building.
- **Disaster risk reduction aims to minimize the vulnerabilities of people to disasters** and can hence be considered a component of an integrated system of social protection in which it helps prepare against, prevent, and mitigate the impact of possible shocks. It is directly related to humanitarian action both as it aims to improve preparedness and response to emergencies and as it is used in early recovery to rebuild better infrastructure and systems to withstand future disasters.
- **Humanitarian assistance can promote resilience** if the preparation of interventions is based on a sound understanding of the crises, so they can as much as possible act on their causes, and if negative possible consequences of emergency responses are fully considered (such as creating dependencies). Conversely, using early recovery approaches in emergencies can help rebuild infrastructure and food production systems more adapted to resist recurrent shocks.

Source: Adapted from Fallavier 2014.

At the same time, lack of government capacity is a real constraint to extending the coverage of existing social protection programs, many of which are public works projects. Such capacity limitations are particularly acute in remote pastoral areas. Successful programs have made a concerted effort to build implementation capacity, including a perennial shortage of staff. Similar challenges relate to the delivery of insurance, which depends upon high-quality, reliable data sources and means of collecting and making cash payments. There may thus be prerequisites in terms of information sources and banking infrastructure that need to be met before insurance can be scaled up in dryland areas.

Recent innovations in delivery mechanisms, particularly the use of ICT, offer opportunities to reach remote populations, which is of particular interest to dryland regions. In northern Kenya, investments in solar panels and smart card technology enabled the Hunger Safety Net Program to create a payment system that is responsive to the mobile lifestyles of pastoral populations. In Somalia, mobile phone technology has played an important role in the Shaqodoon initiative, which uses interactive Somali-language audio programs on financial literacy and entrepreneurship to link youth to employment opportunities via mobile phones and the internet (Lind and Kohnstamm 2014).

Notes

1. A unified registry enables social protection agencies and programs to register their beneficiaries as well as to record and monitor essential data on them and on what they receive. To set up a unified registry, agencies agree on the types of information to collect from the potential beneficiaries and on the method to do so. Elements of a shared management information system can then be developed to include a common registry, but also shared reporting on activities and financial data. They then use similar indicators, forms, and tools to populate a shared database. Data can be used to plan for intervention, and to monitor and evaluate impacts of different approaches. Along with this, a knowledge base can develop on the different programs implemented. This allows to keep track of what programs reach who and when and what beneficiaries received. This helps agencies minimize the duplication of efforts while ensuring that if they operate in similar locations, they use approaches that complement rather than undermine each other's. With time, it helps build a valuable database on vulnerability and on different approaches to respond to it, which can feed research and be invaluable for policy and programmatic decisions.

2. Supporting an integrated approach to climate change adaptation is a good example of how to best develop synergies between different modes of risk management: (i) a component of social protection may strengthen the resilience of vulnerable populations by helping them adapt their modes of livelihoods (a role of protection and promotion, through training in technical skills and the provision of weather-indexed insurance for instance); (ii) a risk-reduction approach would help develop an early warning system, identify vulnerable groups; and reduce their exposure to risk (a role of risk reduction and adaptation to climate change); and (iii) the mechanisms of social protection (targeting, administration, logistics) would support humanitarian actors to distribute the goods and services necessary in a period of crisis or emergency to targeted vulnerable populations on a wide scale.

CHAPTER 7

Financing a National Adaptive Social Protection System

Among African governments, limited financing has long been seen as an impediment to developing integrated systems of social protection, which are perceived as a luxury that poor countries cannot afford. The annual cost of Ethiopia's Productive Safety Net Program is equivalent to 1.2 percent of GDP, while safety net coverage in Kenya is equivalent to 0.80 percent of GDP. In comparison, international evidence suggests that it is possible to achieve national coverage for a target population with a single program for 1–2 percent of GDP. The efforts of these three leaders can provide examples to many of the Sahelian countries, as well as Somalia, Sudan, and South Sudan, whose investments in safety net programs have been modest. With Senegal spending 0.27 percent of its GDP on safety nets, Mali 0.5 percent, and Burkina Faso 0.6 percent, only Mauritania with 1.3 percent is in the 1–2 percent range of the average expenses in poor nations globally.

While safety net programs are often thought to be expensive, studies and simulations have shown otherwise: the cost of not protecting the poor from the effects of shocks can result in even more costly setbacks in terms of long-term development, and no nation can afford not to protect the very poor.[1] Moreover, extending the coverage of existing safety net programs is usually more cost-effective than relying on humanitarian responses or subsidies in times of crisis. For example in Kenya, reorienting existing spending on general food distribution (estimated to cost US$61 million per year) would double the current levels of financing available for cash transfers and make possible high rates of coverage of poor, vulnerable households. In Ethiopia, since the Productive Safety Net Program was launched in 2005, the government has received US$623.6 million per year on average for humanitarian responses, an amount that, if allocated to the Productive Safety Net Program, could extend regular support to a large proportion of the population living below the poverty line. As shown in table 7.1, in Niger, providing regular cash transfer support to 20 percent of the poor population would cost US$102 million per year, as compared with an annual average of US$218 million spent on humanitarian response, on average, between 2010 and 2013. Moreover, a comparative study of the relative costs of late humanitar-

Table 7.1 Cost of Safety Net Support to Poorest Households Compared to Humanitarian Responses

	Annual cost of regular safety net support for 20 percent of poor households in 2015 (US$)					Average cost of humanitarian response, 2010–2013
	Hyper-arid	Arid	Semi-arid	Dry sub-humid	Total	
Burkina Faso	0	1,371,749	88,833,727	11,782,273	101,987,750	48,555,902
Chad	781,398	17,128,141	48,718,180	17,214,163	83,841,882	298,148,319
Mali	210,643	14,643,841	69,557,074	16,788,531	101,200,089	77,423,890
Mauritania	3,107,358	15,568,742	825,661	0	19,501,761	34,784,819
Niger	1,681,344	52,017,414	48,277,168	0	101,975,926	218,221,834
Senegal	0	9,016,207	66,455,931	7,781,703	83,253,841	7,357,294
Total	5,780,743	109,746,094	322,667,740	53,566,670	491,761,248	684,492,057

Note: Number of poor households calculated based on the national poverty line of each country for 2015 population estimates. Annual cost of safety net support estimated to be US$300 per household.

ian responses, early action, and building resilience to disasters in Kenya and Ethiopia provides evidence on the cost of different interventions, presents the most cost-efficient types of actions to support, and finds that early action (prior to the crisis through the use of early warning systems) is far more cost-effective than late humanitarian response (Cabot Venton et al. 2012).

Costing of Safety Net Coverage in Africa's Drylands

National safety net programs may be cost-effective relative to humanitarian responses, but they can still require a significant commitment of resources—with the size of the commitment depending on the scope of coverage and the level of support provided. In a world of unlimited resources and perfect targeting, national safety net programs could theoretically be used to make all drought-vulnerable households living in drylands[2] resilient by providing them with cash transfers in the amounts needed to bring every household up to the poverty line.

Figure 7.1 shows the estimated cost in 2030 of providing safety net support to bring all households vulnerable to drought to the poverty line as a percentage of GDP. The estimated costs range from less than 0.5 percent of GDP in countries with relatively high GDP per capita (for example, Mauritania) to almost 5 percent of GDP in countries with relatively low GDP per capita and extensive dryland populations (for example, Niger).

Policy makers do not live in a world of unlimited resources, and in many dryland countries, by 2030 the cost of providing cash transfer support to drought-vulnerable populations is likely to be unaffordable. But if they are well designed and efficiently administered, national safety net programs need not be prohibitively expensive and can cover a reasonable percentage of the poor population and make them more resilient to droughts. The umbrella model was used to estimate the number of vulnerable people in 2030 who could be made resilient

Figure 7.1 Share of GDP Required to Protect Drought-Vulnerable Population with Safety Net Support, Selected Countries, 2030

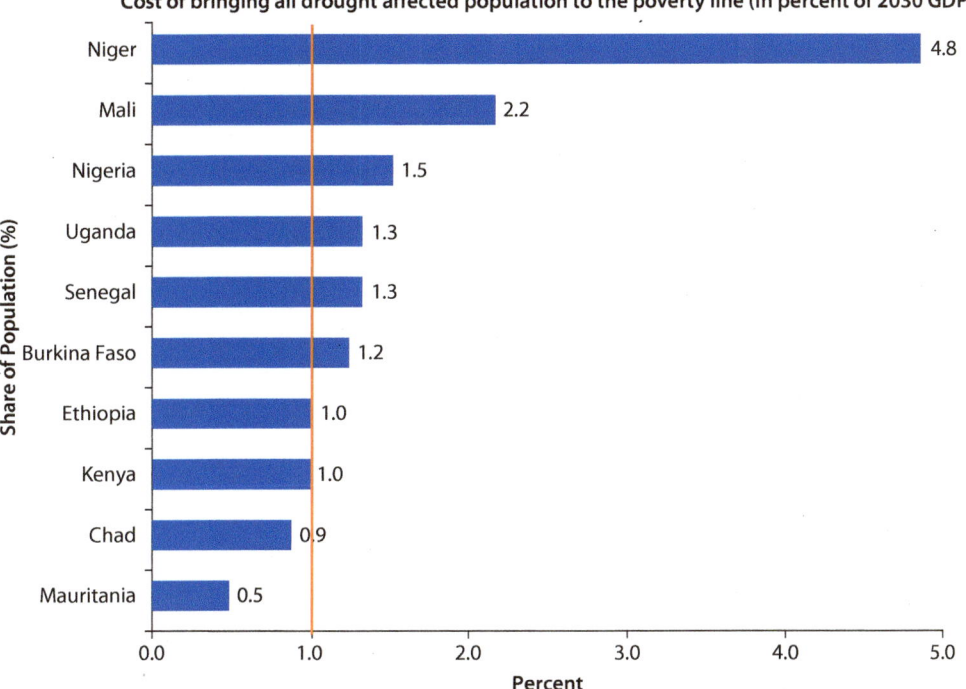

Cost of bringing all drought affected population to the poverty line (in percent of 2030 GDP)

Note: The chart shows the cost (expressed as a percentage of 2030 GDP for drylands, assumed proportional to the share of the population living in drylands) of bringing all drought-affected people to the international poverty line through cash transfers. The cost is calculated taking into account the country-specific depth of poverty, as estimated by the 2010 poverty gap index obtained from the World Bank PovCalnet database. Figures for 2030 GDP are based on the average growth scenario as defined in the umbrella model (Carfagna, Cervigni, and Fallavier 2016).

in the Horn of Africa and the Sahel using some conservative cost estimates. We prepared costs estimates per country for basic safety net packages using population projections for 2030 by aridity zones, and poverty lines of US$1.25 per person per day for all countries except Niger, for which we used detailed poverty breakdowns by aridity zones. We took poverty as a proxy of vulnerability.

We used three instruments:

- An unconditional accompanied cash transfer ("**UCT+**") provided on a monthly basis to the most vulnerable households (mostly women in families with young children, or elders living with grandchildren) and accompanied by an infant nutrition package or other measure adapted to local needs, for a cost per household evaluated at US$300 per year;
- A guaranteed employment scheme of 60 to 90 days of cash-for-work for chronically poor households who have some workforce, at a cost of US$190 per household per year (based on one person working);

- The cost of a weather-based crop or livestock insurance premium ("INS") paid on behalf of vulnerable households who own cultivable land or livestock, at an estimated cost of US$150 per year. (This is an instrument that must still be tested on a larger scale in the region, but has potential, even if only to help households join mutual insurance groups they were previously excluded from.)

With these instruments, we developed a standard package that includes 70 percent of unconditional cash transfers, with the remaining 30 percent split between cash-for-work activities and an insurance premium for an average cost per household of US$261, or approximately CFAF 13,000 per month (see table 7.2).

We used an average household size of six persons, and did not propose a differentiated package for urban versus rural areas or different livelihoods. Note that these are rapid calculations that do not intend to replace more advanced work underway in each country. They were produced mainly to illustrate the relative cost of investing in safety net packages and feed the discussion to develop better safety net packages across the region.

If every country in these two regions were to invest 1 percent of GDP[3] annually in social protection programs, 22.5 million people could be covered in 2030 with a basic social protection program[4] in the Horn of Africa (approximately 71 percent of the projected vulnerable population in that subregion), and 45 million people could be made resilient in the Sahelian countries (approximately 65 percent of the vulnerable population in that subregion). Figure 7.2 shows that countries with a higher level of GDP per capita can achieve a higher level of coverage than countries with a lower level of GDP per capita.

The umbrella modeling is also used to estimate the cost of a basic scalable social protection program that can cover 25 percent of the poor and vulnerable population in a normal year, or 35 percent of the vulnerable in case of a mild drought, 50 percent for a moderate drought, and 65 percent for a severe drought (see table A.2 for a correspondence between the percentage of people covered and the income cut-off used in the umbrella model). The estimated amount of the transfers also depends on the type of package provided.[5] In West and East

Table 7.2 Per Household Cost of Safety Net Response Under Different Crisis Scenarios

Packages per crisis scenario	Share of instrument in package			Cost per package		
	UCT+ (%)	CFW (%)	INS (%)	Initial	Additional cost per year (%)	Total
In a regular year as part of permanent safety nets	70	15	15	$261	20	$313
In a year of mild drought	50	35	15	$239	30	$311
In a year of medium drought	35	50	15	$223	30	$289
In a year of severe drought	15	70	15	$201	35	$271

Note: UCT = unconditional cash transfer; CFW = cash-for-work; INS = insurance premium.

Figure 7.2 Percent of Vulnerable Population Covered in 2030 with 1 Percent of GDP

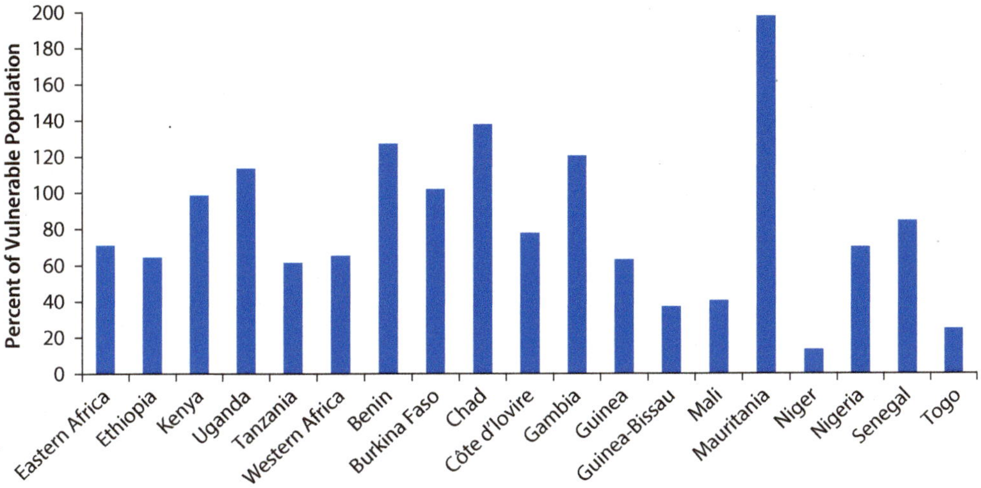

Note: Vulnerable population includes those below the poverty line in agriculture in the drylands. The GDP figures are prorated for the population in agriculture in the drylands.

Africa, covering 25 percent of the estimated 101 million vulnerable in a regular year in 2030 would cost 0.3 percent of GDP. In case of mild, moderate, and severe drought the cost would go up to 0.5, 0.76, and 1 percent of GDP, respectively. Figure 7.3 shows the estimates for each of the countries included. It is clear that poorer and vulnerable countries like Niger will need a proportionally larger amount of resources and therefore may be more in need of foreign aid, especially in a drought situation.

A relatively clear message comes from the projections. A minimum level of social safety nets—which is essential to protect the poorest from chronic stress and from sudden climate-related shocks—is affordable. Social safety nets are a low-cost approach to not only help bridge a revenue gap for the poorest, but also impart them with human and physical capital to strengthen their livelihoods against crises (through the training that comes along with the cash transfers, the climate-resilient productive infrastructure built through cash-for-work, and the incentive to diversify from the agriculture-insurance component).

Of course, when a shock has occurred and people are suffering, political and humanitarian considerations will almost always demand additional funds for short-term *ad hoc* transfers to cover those affected. However, emergency programs are not as effective at building long-term resilience. The challenge for policy makers is to strike an appropriate balance between permanent and emergency programs, a task made especially difficult by the fact that financing needs for safety nets are inherently unpredictable. The emerging experience with scalable safety nets suggests that investments made in permanent systems reduce the costs associated with delivery support to households negatively affected by drought.

Figure 7.3 Cost of Covering Vulnerable Population in 2030 in a Regular Year or in the Case of Mild, Moderate, and Severe Drought in Percentage of GDP

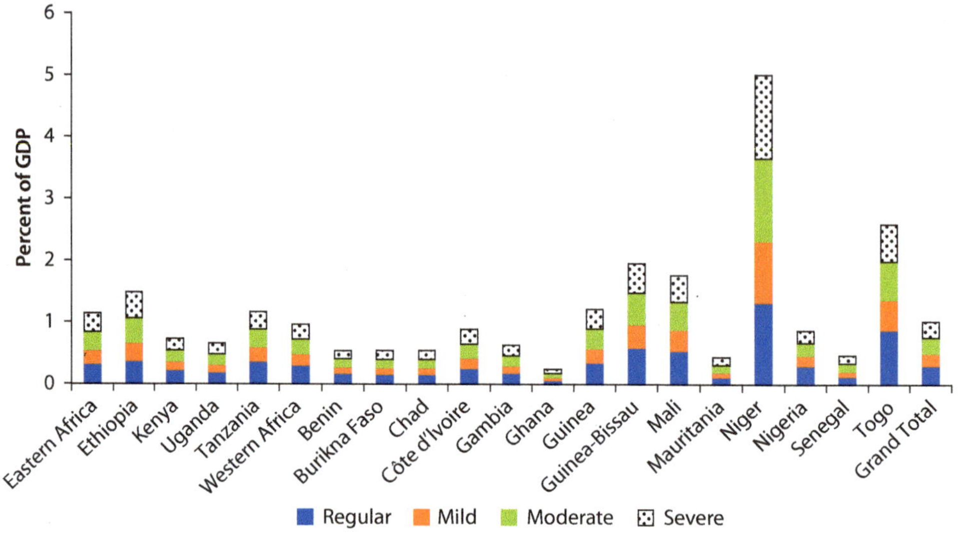

Source: Calculations using data produced by the umbrella model (Carfagna, Cervigni, and Fallavier 2016).

Beyond technical approaches to better manage risk, reducing exposure and sensitivity to shocks in the long term further requires designing development programs that build the resilience of vulnerable populations to crises and enable them to break out of poverty cycles. This includes diversifying livelihoods bases, which itself requires ambitious investments in functional education and market-able skills, as well as the development of climate-adaptive cultivation techniques and marketing and transportation infrastructure, but it also includes urgently finding ways to shorten the demographic transition and to address the bottle-necks that keep a very large part of the population extremely vulnerable to preventable causes of morbidity and death, including the wide gender gap in accessing basic services.[6] This can only happen with strong political engagement from government.

Notes

1. See for example the Africa Social Protection Strategy (World Bank 2012a).

2. Here, all drought-affected households is defined as all vulnerable households that are unable to cope in a regular year.

3. These funds are prorated for dryland populations that depend on agriculture. In Ethiopia for example in 2010, 47 percent of population lived in drylands and 70 percent of them were engaged in agricultural activities. This means that a little more than 30 percent of the total population was covered, so it would be 31 percent of 1 percent of GDP that is allocated to dryland areas in a regular year—and allocated only to poor agricultural producers in the drylands.

4. A per capita transfer of US$60 per person, used in this estimate, includes the cost of basic coverage equal to US$261 per household plus 15 percent for administration fees and 20 percent for leakage to the non-poor.

5. The package includes a varying combination of cash transfers, public works programs, and insurance support equal to US$313 per household in a regular year, US$311 in mild drought, US$289 in moderate drought, and US$271 in severe drought. The cost of those packages includes administrative expenses and leakages to the non-poor.

6. A 2013 World Bank assessment of vulnerability points to better tapping the potential for developing the pastoralist economy in the north, and developing new related industries in urban centers. It argues for the scaling up of livestock production and processing for the domestic market both to support the economy and to help feed the growing urban centers. Taking this road would require large investments in agricultural extension services, but also in transportation infrastructure and in vocational training in the processing of dairy, meat, and other animal products to support the creation of value added (World Bank 2013b).

CHAPTER 8

Conclusion

Social protection programs will be a key component of strategies to increase resilience and reduce vulnerability in the drylands. If present trends continue, by 2030 dryland regions of East and West Africa will be home to an estimated 429 million people, up to 24 percent of whom will be living in chronic poverty. Many others will depend on livelihood strategies that are sensitive to the shocks that will hit the region with increasing frequency and severity, making them vulnerable to falling into transient poverty. Social protection programs thus will be needed in the drylands to provide support to those unable to meet their basic needs. Some of these people will require long-term support, while others will require periodic short-term support because of income losses due to shocks (for example, crop failure following a drought) or as a result of lifecycle changes (for example, loss of a breadwinner).

Safety net programs can increase resilience in the short term by improving the coping capacity of vulnerable households. Rapidly scalable safety nets that provide cash, food, or other resources to shock-affected households can allow them to recover from unexpected shocks. Scaling up an existing safety net program can be far less expensive than relying on appeals for humanitarian assistance to meet urgent needs. Despite the fact that safety nets are a more effective response to poverty and vulnerability than emergency assistance, funding for safety nets is low and flows of humanitarian resources to countries in the Horn of Africa and the Sahel remain high.

Social protection programs can increase resilience over the longer term by reducing vulnerable households' sensitivity to shocks especially if combined with other development programs. Safety net programs must be complemented by other types of social protection programs that enable chronically poor households to build their productive assets and expand their income-earning opportunities. Providing predictable support to chronically poor households and enabling them to invest in productive assets and access basic social services can effectively reduce their sensitivity to future shocks, help them to participate in the growth process, and take advantage of the investments made in agriculture and pastoralist activities proposed in the drylands. Households covered by well-functioning social protection programs are less likely to resort to negative coping strategies, such as pulling their children out of school and selling productive assets.

The dynamic nature of vulnerability in dryland areas draws attention to the need for safety net programs to be able to scale up in the face of shocks and to scale back down when these pass. In dryland areas, such instruments may be even more important than in non-dryland areas given the levels of vulnerability and exposure to shocks. Emergency support should be provided on an occasional basis whenever a set of pre-defined triggers is met and in a manner that complements, rather than replaces, the support extended through scalable safety nets. Effective early warning and monitoring systems are needed to alert policy makers and guide the response.

Social protection programs must be tailored to address the unique circumstances of dryland populations. The needs of poor households living in drylands often differ from those of poor households living in more favorable environments or in urban areas. For this reason, one-size-fits-all programs implemented at the national level often fail to adequately address the needs of dryland populations. Interventions designed to strengthen the livelihoods strategies of dryland populations and build their resilience will not be effective if they fail to account for their specific needs. Program delivery mechanisms similarly need to respond to the specific needs of dryland populations (for example, by accommodating the mobility of pastoral populations).

Capacity constraints will need to be overcome to ensure effective implementation of social protection programs in the drylands. Effective implementation of social protection programs in the drylands is made difficult by the limited presence of public agencies and a lack of infrastructure. Incentives are needed to attract and retain qualified staff in hardship posts. Investments in transportation systems and information technology are needed to improve mobility and reduce transactions costs associated with implementing social protection programs in remote dryland areas.

APPENDIX A

Table A.1 Vulnerability to Different Drought Scenarios, 2030

People unable to cope in:	2010 Normal year	2030 Normal year	Mild drought	Medium drought	Severe drought
Eastern Africa	25,182	31,810	38,762	44,877	50,031
Ethiopia	9,956	12,044	15,498	18,677	21,401
Kenya	3,720	4,500	5,301	6,021	6,661
Uganda	1,791	1,996	2,420	2,797	3,126
Tanzania	9,715	13,269	15,542	17,382	18,843
Western Africa	42,216	69,528	81,719	92,112	100,805
Benin	1,071	800	925	1,031	1,119
Burkina Faso	5,530	5,551	6,703	7,705	8,564
Chad	2,799	3,995	4,760	5,475	6,133
Côte d'Ivoire	819	1,246	1,491	1,720	1,930
Gambia	370	555	666	769	864
Ghana	844	456	570	682	785
Guinea	171	216	262	302	337
Guinea-Bissau	22	35	42	48	52
Mali	3,572	5,484	6,501	7,337	8,005
Mauritania	447	602	768	927	1,069
Niger	4,412	15,179	19,284	22,791	25,688
Nigeria	19,119	29,897	33,208	35,849	37,946
Senegal	1,949	3,510	4,289	5,018	5,681
Togo	1,091	2,004	2,250	2,458	2,631
Grand Total	67,398	101,338	120,481	136,989	150,836

Source: Carfagna, Cervigni, and Fallavier (2016) under scenarios of medium GDP growth and medium fertility rates.

Table A.2 Share of Vulnerable Population to be Covered under Different Scenarios

Income cut off	Agricultural population with daily income under			
	$1.25	$1.44	$1.63	$1.81
Regular year	25%			
Mild drought year	25%	35%		
Medium drought year	25%	35%	50%	
Severe drought year	25%	35%	50%	65%

Note: Liberia and Sierra Leone are excluded from the calculations since they do not have drylands. Djibouti, Eritrea, Somalia, South Sudan, and Sudan are also excluded as historical data are lacking (on poverty and GDP to project them to 2030).

Bibliography

Béné, C. 2011. "Social Protection and Climate Change." *IDS Bulletin* 42 (6): 4.

Béné, C. 2012. Social Protection and Resilience to Climate and Disaster *IDS Programme Briefing*. Institute of Development Studies, World Bank, Centre for Social Protection, UK Aid, UN Economic Commission for Africa, 4.

Cabot Venton, C. 2011. The Benefits of a Child-Centred Approach to Climate Change Adaptation. UNICEF, Plan International, 28.

Cabot Venton, C., C. Fitzgibbon, T. Shitarek, L. Coulter and O. Dooley. 2012. *The Economics of Early Response and Disaster Resilience: Lessons from Kenya and Ethiopia*. London: Department for International Development (DFID), 84.

Carfagna, F., R. Cervigni and P. Fallavier, editors. 2016 (forthcoming). *Mitigating Drought Impacts in Drylands: Quantifying the Potential for Strengthening Crop- and Livestock-Based Livelihoods*. World Bank Studies. Washington, DC: World Bank.

Cervigni, Raffaello, and Michael Morris, editors. 2016. *Confronting Drought in Africa's Drylands: Opportunities for Enhancing Resilience*. African Development Forum series. Washington, DC: World Bank.

Chantarat, S., A. G. Mude, C. B. Barrett, and M. R. Carter. 2013. "Designing Index-Based Livestock Insurance for Managing Asset Risk in Northern Kenya." *Journal of Risk and Insurance* 80 (1): 205–37. doi:10.1111/j.1539-6975.2012.01463.x

Comité du Pilotage de la Stratégie Nationale de Protection Sociale. 2012. STRATEGIE NATIONALE DE PROTECTION SOCIALE EN MAURITANIE—Elément essentiel pour l'équité et la lutte contre la pauvreté. Ministère des Affaires Economiques et du Développement, avec l'appui de l'UNICEF, 148.

Davies, M., and J. Leavy. 2007. Connecting Social Protection and Climate Change Adaptation. Institute of Development Studies, University of Sussex, 2.

Davies, M., K. Oswald, T. Mitchell, and T. Tanner. 2008. Climate Change Adaptation, Disaster Risk Reduction and Social Protection: Centre for Social Protection, Climate Change and Development Centre, Institute of Development Studies, University of Sussex.

Davies, M., K. Oswald and T. Mitchel. 2009. "Climate Change Adaptation, Disaster Risk Reduction and Social Protection." In *Promoting Pro-Poor Growth: Social Protection*, 201–17. Paris: Organisation for Economic Co-operation and Development.

Davies, M., C. Béné, A. Arnall, T. Tanner, A. Newsham, and C. Coirolo. 2013. "Promoting Resilient Livelihoods through Adaptive Social Protection: Lessons from 124 programmes in South Asia." *Development Policy Review* 31 (1): 27–58.

Fallavier, P. 2013a. Linking Social Protection, Humanitarian Action, and Disaster Risk Reduction: Review of the Literature and Mapping of Best Practices. Prepared for the Division of Policy and Strategy, and the Office of Emergency Programmes, United Nations Children's Fund (UNICEF), 107.

Fallavier, P. 2013b. Protection Sociale et Résilience au Niger: Liens existants et à renforcer pour mieux protéger et promouvoir. Note politique préparée pour la Banque Mondiale et la Cellule Filets Sociaux, Cabinet du Premier Ministre, République du Niger en collaboration avec UNICEF, 88.

Fallavier, P. 2014. *Social Protection in Sahel's drylands: Strengthening Resilience and Promoting Opportunities While Contributing to Equitable Growth*. Background report prepared for the World Bank. World Bank, Washington, DC.

Frankenberger, T., T. Spangler, S. Nelson and M. Langworthy. 2012. Enhancing Resilience to Food Security Shocks in Africa Discussion Paper prepared for DFID, the FAO, The New Partnership for Africa's Development of the African Union, USAID, and the World Bank: TANGO International.

High Level Panel of Experts on Food Security and Nutrition. 2012. Food Security and Climate Change *HLPE Reports*. Vol. 3. Rome: Committee on World Food Security, High Level Panel of Experts on Food Security and Nutrition, 98.

Hoddinott, J., and J. Lind. 2013. "The Implementation of the Productive Safety Nets Programme in Afar and Somali Regions, Ethiopia: Lowlands Programme Outcomes Report." International Food Policy Research Institute, Washington DC.

Kuriakose, A.T., R. Heltberg, W. Wiseman, C. Costella, R. Cipryk, and S. Cornelius. 2013. "Climate-Responsive Social Protection." *Development Policy Review* 31: 19–34. doi:10.1111/dpr.12037.

Lind, J. 2013. Review of Social Protection Programmes and Projects in the IGAD Region: Institute of Development Studies, University of Sussex, 17, + annexes.

Lind, J., and S. Kohnstamm. 2014. *Review of Social Protection Programmes and Projects in the IGAD Region*. Background report prepared for the World Bank. Washington, DC: World Bank.

Ministry of State for Planning National Development and Vision 2030. 2012. Kenya Social Protection Sector Review. Republic of Kenya, 149.

Monchuk, V. 2014. *Reducing Poverty and Investing in People: The New Role of Safety Nets in Africa*. Washington, DC: The World Bank.

Newsham, A., M. Davies, and C. Béné. 2011. *Making Social Protection Work for Pro-Poor Disaster Risk Reduction and Climate Change Adaptation*. Addis Ababa, Ethiopia: Institute of Development Studies, 22.

Ovadiya, M., and C. Costella. 2013. *Building Resilience to Disaster and Climate Change Through Social Protection*. Washington, DC: Rapid Social Response, Global Facility for Disaster, 36.

Uganda Ministry of Gender, Labour and Social Development. 2014. Social Protection Sector Review. Draft September 2014. Kampala: Ministry of Gender, Labour and Social Development.

United Nations. 2011. Global Drylands: A UN System-Wide Response. The Environment Management Group, United Nations, 131.

United States Social Security Administration. 2013. *Social Security Programs Throughout the World: Africa*. Washington, DC: Office of Research and Statistics, United States Social Security Administration; and International Social Security Association.

UNOCHA. 2015. Sahel: Food Insecurity 2011–2015 | ReliefWeb. http://reliefweb.int/disaster/ot-2011-000205-ner

Wernerman, J. 2012. Resilience Presentation Given at UNICEF's Workshop on Cash Transfers in Emergency/for Resilience Workshop, Held November 28–30, 2012. Dakar: UNICEF, West and Central Africa Office, 5 slides.

World Bank. 2004. *Burkina Faso—Risk and Vulnerability Assessment*. In edited by K. Subbarao and M. Temourov. Washington, DC: Human Development Unit, AFTH2, Africa Region, The World Bank, 77.

World Bank. 2006. *Senegal—Managing Risks in Rural Senegal: A Multi-Sectoral Review of Efforts to Reduce Vulnerability*. In edited by M. Temourov. Washington, DC: Human Development II (AFTH2), Africa Region, The World Bank, 124.

World Bank. 2009a. MALI—Social Safety Nets (A. R. Human Development, Social Protection, Trans.) Washington, DC: World Bank, 115.

World Bank. 2009b. NIGER Food Security and Safety Nets. Human Development, AFTH2, Country Department AFCF1, Africa Region, The World Bank, 83.

World Bank. 2010. *World Development Report 2010: Development and Climate Change*. Washington, DC: The World Bank.

World Bank. 2011a. *Burkina Faso—Social Safety Nets*. Washington, DC: Human Development Department, Social Protection Unit, Africa Region, The World Bank, 121.

World Bank. 2011b. *Mali—Social Safety Nets*. Washington, DC: Human Development Department, Social Protection Unit, Africa Region, The World Bank, 169.

World Bank. 2011c. *NIGER: Investing for Prosperity—A Poverty Assessment*. Washington, DC: Poverty Reduction and Economic Management, Africa Region, The World Bank, 132.

World Bank. 2012a. *Managing Risk, Promoting Growth: Developing Systems for Social Protection in Africa The World Bank's Africa Social Protection Strategy 2012–2022*. Washington, DC: Human Development Africa, The World Bank, 77.

World Bank. 2012b. *The World Bank 2012–2022 Social Protection and Labor Strategy: Resilience, Equity, and Opportunity*. Washington, DC: The World Bank, 109.

World Bank. 2013a. *Republic of Senegal—Senegal SP—Safety Net Assessment*. Washington, DC: Africa Social Protection Group for West and Central Africa (AFTSW), the World Bank, 66.

World Bank. 2013b. *Stresses in the Sahel Region: Risk Vulnerability Analysis—A Case Study of Mali*. Washington, DC: Africa Region Conflict and Social Development Unit (AFTCS), the World Bank, 37.

World Bank. 2014. Project Appraisal Document on a Proposed Credit in the amount of SDR391.9 million to the Federal Democratic Republic of Ethiopia for a Productive Safety Nets Project 4. Washington, DC: The World Bank.

World Bank. 2015. "Opportunities for Social Protection to Address Poverty and Vulnerability in a Crisis Context." Mali Social Protection Policy Note. Africa Social Protection, World Bank, Washington, DC, 51.

ECO-AUDIT

Environmental Benefits Statement

The World Bank Group is committed to reducing its environmental footprint. In support of this commitment, the Publishing and Knowledge Division leverages electronic publishing options and print-on-demand technology, which is located in regional hubs worldwide. Together, these initiatives enable print runs to be lowered and shipping distances decreased, resulting in reduced paper consumption, chemical use, greenhouse gas emissions, and waste.

The Publishing and Knowledge Division follows the recommended standards for paper use set by the Green Press Initiative. The majority of our books are printed on Forest Stewardship Council (FSC)–certified paper, with nearly all containing 50–100 percent recycled content. The recycled fiber in our book paper is either unbleached or bleached using totally chlorine free (TCF), processed chlorine free (PCF), or enhanced elemental chlorine free (EECF) processes.

More information about the Bank's environmental philosophy can be found at http://www.worldbank.org/corporateresponsibility.